ECLECTIC COLLECTIONS

室内软装藏品

（澳）史蒂芬·克拉弗缇　编著

于丽红　译

广西师范大学出版社　images
· 桂林 ·　　　　　　Publishing

图书在版编目(CIP)数据

室内软装藏品/(澳)克拉芙缇 编;于丽红 译.—桂林:广西师范大学出版社,2016.2
ISBN 978－7－5495－7830－6

Ⅰ.①室… Ⅱ.①克… ②于… Ⅲ.①室内装饰设计-图集 ②收藏-世界-图集 Ⅳ.①TU238－64 ②G894－64

中国版本图书馆 CIP 数据核字(2016)第 001628 号

出 品 人:刘广汉
责任编辑:肖　莉　于丽红
版式设计:吴　迪

广西师范大学出版社出版发行

(广西桂林市中华路22号　邮政编码:541001)
(网址:http://www.bbtpress.com)

出版人:何林夏
全国新华书店经销
销售热线:021－31260822－882/883
利丰雅高印刷(深圳)有限公司印刷
(深圳市南山区南光路1号　邮政编码:518054)
开本:787mm×1 092mm　1/12
印张:$18\frac{2}{3}$　　字数:30 千字
2016 年 2 月第 1 版　　2016 年 2 月第 1 次印刷
定价:248.00 元

如发现印装质量问题,影响阅读,请与印刷单位联系调换。

目录
CONTENTS
ECLECTIC COLLECTIONS

4　前言

8　阿利·艾特金 / 动物标本
18　佚名 / 家具：传奇的舒利姆·科瑞姆珀（Schulim Krimper）
28　迈克尔·巴克斯顿 / 当代澳大利亚艺术品
38　克里斯·康奈尔 / 椅子
46　苏珊·柯蒂斯 / 精美艺术品
56　贝亚特和范恩·费希尔 / 精美艺术品
66　桑迪·盖耶 / 当代珠宝
76　克里斯多夫·格拉夫 / 腹语式木偶
86　莎拉·格斯特 / 澳大利亚木箱
96　诺拉·哈格里夫斯 / 珍稀旧时织品
106　约翰·亨利 / 名品家具
116　马丁·希斯科克 / 珍贵贝壳
126　马乔里·约翰斯顿 / 旧时围巾
136　罗谢尔·金 / 瓷器
146　扬尼·劳福德·索尔提斯 / 现当代时装
154　里昂住宅博物馆 / 澳大利亚当代艺术品
164　迈克尔·马丁 / 书籍珍本
174　扬和 莫里斯·米德 / 克拉丽丝·克里夫陶瓷
184　菲利斯·莫菲 / 墙纸
194　苏西·斯坦福 / 蛋糕装饰
204　艾莉森·沃特斯 / 时尚帽子

214　版权许可
216　致谢

前言

史蒂芬·克拉弗缇

INTRODUCTION

STEPHEN CRAFTI

数世纪以来，人们一直都在收藏物品。无论是奇珍异宝，还是引起童年回忆的简单物件，收藏都可以将一个单纯的爱好变成终身使命。尽管在开始收藏的时候，这只是一种适度的兴趣，而如果不加以控制，最终却能变成一种痴迷——经常收藏下一个更加稀奇的物件。

书中的收藏家，收藏行为并不是从他们现有收藏品开始的。开始收藏的时候，可能走了弯路而进入了古玩市场或者二手店。在客户莎拉（SARAH）的案例中，尽管最初她只是为了女儿寻找一个衣柜，并找到一个在20世纪60年代由一位学徒制作的非常简易的木箱，但最终她成为了澳大利亚小木箱的收藏家。一个箱子接着一个箱子，现在她的收藏品已经超过了200个木箱。

同样，建筑师菲利斯·莫菲（PHYLLIS MURPHY）的藏品在20世纪80年代改变了方向。她和同样为建筑师的丈夫，约翰·莫菲（JOHN MURPHY）决定搬到乡下。他们发现了一个古老的油漆装饰的小屋，里面装了数百卷壁纸。这些壁纸甚至可追溯到19世纪早期，这引发了她对古老墙纸的兴趣。30年后，莫菲已经收集了——尽管不是最多的——一个时代的知名墙纸。每种墙纸都细心的加上标签，收进目录。莫菲继续寻找墙纸，很多人认为自己的墙纸很有趣而直接联系她。其他收藏家，例如室内设计师桑迪·盖耶（SANDY GEYER），收藏了大量的当代珠宝。她的首饰盒中充满了有趣的饰品。一些饰品，例如晚期澳籍日裔艺术家毛利·船木（MARI FUNAKI）的珠宝，被完美的展示在盖耶的起居室里的有机玻璃箱中。盖耶定期的打开箱盖，选择一枚胸针（部分雕刻，部分珠宝）配戴。和本书中的很多重要收藏家一样，盖耶一直在收藏20世纪60年代晚期的当代珠宝，那时很多人都佩戴珍珠穿运动服。

艺术品收藏家哈利和苏珊·柯蒂斯（HARRY AND SUSAN CURTIS）开始时也是着眼于20世纪60年代的当代艺术品,和现在很多艺术家在开始时都是谨慎选择从事一种艺术品收藏的做法不同的是，哈利和苏珊·柯蒂斯是之后在偶然的机会下碰到了一种独特的蚀刻版画。这对刚结婚的年轻夫妇推迟了为新家购置生活必需品，例如拖把或者扫帚，因

为他们在墙上的蚀刻版画中获得了更大的快乐。不必惊讶，这次的蚀刻版画使他们开始研究当时一些其他同时代的伟大艺术家。他们看见了什么其他人没有注意到的？贝亚特和范恩·费希尔（BEATA AND VANN FISHER）也发现了收藏当代艺术品的重要性。他们现在已经有了一些非凡的收藏品，包括名画、雕塑和艺术品，这些藏品完美的融入到了他们的家中。倾斜的墙和门允许艺术品重新悬挂，创造了一个动感的千变万化的收藏品展示区。

另外一位收藏家，科比特·里昂（CORBETT LYON）为了和公众分享他的当代澳洲艺术品收藏，他将自己的家变成了一个住宅博物馆。多年以来，迈克尔·巴克斯顿（MICHAEL BUXTON）的艺术收藏品也一直在增加，目前收藏品数量已经可以和国家最大最重要的当代艺术品收藏馆相媲美。

扬尼·劳福德·索尔提斯（JANNI LAWFORD SOLTYS）是从一个不同的方向开始收藏的。作为20世纪60年代和70年代的时尚模特，索尔提斯被伦敦的跳蚤市场和二手旧时服装店所吸引。几十年后，她的衣柜和木箱中已经充满了20世纪早期一直到近代的设计师专门设计的时装。直到日本时尚设计师在20世纪80年代早期榜上有名，索尔提斯才意识到这些服装的真正价值。

珍稀贝壳、书籍和名画也在本书中占据了重要位置，而且也有一些收藏品很难估价。苏西·斯坦福（SUZIE

STANFORD)的蛋糕装饰收藏开始于她父母的结婚蛋糕。很快，斯坦福开始不断地寻找一个接一个的蛋糕装饰。

本书不单收集了大量精美而珍贵的收藏品，并且通过收藏品讲述了丰富迷人的故事。本书也发现了人们用多久才能找到下一位收藏家的"关注"。一位收藏家，专注于一位20世纪中后期的家具设计师的作品，他的收藏爱好是不会停止的。他家里的每个房间都充满了家具，在家中放不下的，会放到仓库里。没有什么比这位收藏家接到某个电话，告知他有这位设计师的一件家具要卖的消息更让他快乐的了。当被问到为什么拥有下一件家具如此重要的时候，从某种程度上讲，这是一种无法解释的忧虑。

收藏是成瘾的、着迷的，你可以说这很难被完全的解释。但等到能够解释的时候，就已经晚了——一个人的住处已经充满了名副其实的非凡收藏品。

动物标本
阿利·艾特金

TAXIDERMY
ALY AITKEN

ARTIST ALY AITKEN lives in a multilevel warehouse in the city. Once a workshop for motorbikes, it's now a sanctuary for Aitken's family. With her workshop at ground level and studio at first level, Aitken is continually walking through her beautifully curated spaces. 'I don't consider myself a collector. But taxidermy is integral to this place,' says Aitken.

Crows, a hawk, a magpie and fox heads appear intermittently in Aitken's home. In her studio, there are also tails and animal legs, teeth, jaw sections, bones and wings that will form part of her own sculptures. Crows appear in a number of rooms, including three perched on a timber beam in the stairwell. A harrier attached to the stairwell wall appears to take flight. 'It's that experience of flying which first attracted me to the birds,' Aitken admits. 'With the crows, I love seeing their turned-up wings when in flight. It's not dissimilar to watching a dancer, with their upturned hands.'

Aitken's first crow came from a secondhand shop. She removed its timber stand to 'free it up' and refashioned it by attaching it to a falconer's glove. While this crow was relatively inexpensive, Aitken soon discovered why. 'Part of it had been eaten by insects,' she says. Since her first acquisition, Aitken has learnt to regularly spray her taxidermy with insect spray, without oversaturating the piece, and placing it in a garbage bag to absorb the spray. 'It's the best way of fumigating insects,' she says.

While the crows (now a total of six) are comfortable on the rafters, other animals have joined the group. A hare, found at another secondhand furniture store, sits atop a pile of magazines in Aitken's bedroom. The foxes came next – one that is now sitting on a tattered armchair in a room referred to as the 'curiosity cabinet', the other reclined on the floor beside it. Aitken found the first fox at a deer farm, where they also sold taxidermy deer and antlers. 'I don't think of them as taxidermy. I love these animals,' says Aitken, who feels that the modern world has become slightly divorced from nature. 'You often see the landscape through glass windows, in a safe and sealed environment. The taxidermy is a reminder of what we've lost.'

Among the collection of taxidermy are three chickens, originally wandering the back garden of Aitken's home in the suburbs.

TAXIDERMY

There's Gumboot, Pigeon and Robert, all of whom were personally taken to a taxidermist by Aitken once they had died. 'It's not about the memories of the chickens once walking around our garden. It's more they were once alive and can still bring pleasure to this household. Back then I placed them in the living room. They could have been eating seed off the floor. Here, they have found a place on shelves,' says Aitken. And while Aitken isn't a taxidermist, she appreciates the talents of one who is experienced. 'They need to capture a natural pose for it to be successful,' says Aitken.

Some pieces, such as the hares and chickens are displayed in the open – on shelves or rafters. Others, such as one of the magpies, are displayed under a glass dome on the living room level. The backdrop to the magpie is a series of what Aitken describes as 'tacky' postcards found on her overseas travels. The cardboard buildings, representing manufacturing, provide a sharp juxtaposition to the magpie. 'Many of my settings are about contrasts,' says Aitken. While most birds in Aitken's home are taxidermy, she also has a fascinating bird in a cage by American artist

Troy Abbott. The cage was found by Aitken in a secondhand shop and sent to Abbott. But rather than inserting a predictably stuffed bird, he returned the cage with a video installation of a canary – as real as any bird. 'My grandfather used to breed singing canaries,' says Aitken. Other birds, like the large flamingo positioned at the entrance to the living room, are not taxidermy. The painted steel bird sits on a set of wheels, allowing it to be easily moved into different areas of the home.

While some of the taxidermy in the warehouse is presented in its entirety, other pieces are only seen in parts (particularly in Aitken's studio). A fox head, for example, is the crown for a 1920s child mannequin. 'Some of these things are entirely worthless to anyone but myself,' says Aitken, who only purchases taxidermy of non-protected or introduced species that are considered pests by farmers and such.

In the end, it's the sense of surprise that makes Aitken's taxidermy collection so delightful. 'When you walk in, it's not something you expect to see inside, particularly in a warehouse in the city,' adds Aitken.

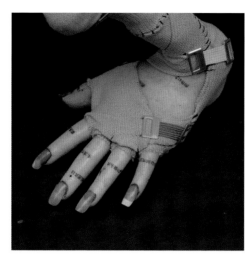

TAXIDERMY

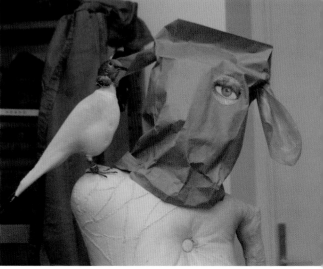

TAXIDERMY

家具：传奇的
舒利姆·科瑞姆珀 （Schulim Krimper）

佚名

FURNITURE
ANONYMOUS
The legendary Schulim Krimper

THE OWNER OF this extraordinary collection of furniture by furniture maker Schulim Krimper has filled their entire home with the designer's pieces. In every room there are sideboards, dining tables, chairs, coffee tables or cabinetry. There's even built-in joinery by Krimper, designed for the owner's parents.

If you look under tables or protective cloths, you'll find more Krimper, who made his impact on the Melbourne design scene from the 1940s through to the 1960s. Originally from the Austro-Hungarian Empire, Krimper designed entire interiors for his Melbourne clientele shortly after he arrived in 1939.

The owner recalls how Krimper would arrive in his dustcoat and take extensive details about what was needed. 'He wore gloves, particularly when a certain built-in unit would arrive. And doorways were always extremely well protected,' says the owner, who started collecting Krimper furniture from the mid-1950s.

Some of the furniture, such as the six-seater lounge suite in the formal dining area and the lamp beside it, belonged to the owner's parents. But these whetted the owner's appetite to collect their own Krimper pieces. In most of the rooms there are dining tables and chairs created by the maker. For example, in the kitchen there are two full dining suites. Buffets, one of the owner's favourite furniture types, can be found in almost every room, including the bedrooms.

Difficult to pigeon-hole, the owner describes Krimper's style as a fusion of various influences, from Biedermeier to folk-like. 'It's quite organic in feel, but it always celebrates the timber, whether it's Queensland walnut or teak.' The depth and breadth of Krimper's designs are also impressive. There are tea trolleys, glory boxes, bedheads and even fine timber bowls and utensils that came as part of a buffet. Some of the pieces recall another time. With the advent of the mobile phone, Krimper's telephone table now seems an oddity.

Other Krimper designs show a fascination for the East. Some of the sideboards and bedheads are inspired by Japanese pergolas.

'Buffets show Krimper's immense talent for joinery,' says the owner, pointing out the drawers with a dovetail-end and the fine brass rods used to strengthen some of the heavier cupboard doors.

The owner's first Krimper purchases were a 1960s bookcase made from Queensland walnut and a daybed from the same period. 'There's no French polishing. Krimper used to wax his furniture. It feels like velvet.' Then there are the more unusual pieces, such as a 1960s cocktail cabinet. One of the rarest pieces in the collection is a unit custom-designed for the first television set. 'Krimper had an exhibition in New York and I had this sent over.'

FURNITURE

FURNITURE

Other pieces feature fine parquetry tops and concealed nooks. But irrespective of the item, they all show a highly developed talent and understanding of engineering, allowing even the heaviest drawers to slide with great ease.

Although having hundreds of Krimper items, the owner hasn't slowed down their desire to collect more. People hear about this collection through word of mouth, and rather than pass on their Krimper to a dealer, they contact this owner to see if they're interested. Some pieces are also found at auctions or in stores. However, the owner can see straight away whether a piece is highly collectable. There's a rare chess table in the billiard room or the highly desirable office desk and chair in the main bedroom – with its chunky glass table top.

While each of the Krimper pieces are treasured, so are the memories of meeting the designer and seeing him deliver pieces to the family home. 'Sometimes he would meet with a client, take down all the details and go away. You wouldn't hear from him for two years and there'd be a phone call announcing it was ready,' says the owner. 'It was never what you had expected, well beyond your expectations.'

FURNITURE

当代澳大利亚艺术品

迈克尔·巴克斯顿

CONTEMPORARY AUSTRALIAN ART

MICHAEL BUXTON

MICHAEL BUXTON, executive director of MAB Corporation, recently bequeathed a significant collection of contemporary Australian art to the University of Melbourne. This forms just one of his art collections acquired over the last couple of decades. There is also Buxton's collection of contemporary American art in Texas – collected with business partner, Brad Nelson. And then there's Buxton's collection of European art that surrounds him at home. And of course, there's his collection of Australian contemporary art displayed in almost every room at MAB's headquarters in Melbourne.

While the depth and range of contemporary art collected by Buxton is exemplary, his interest in art only started in 1974. The painting by the late artist Jeffrey Smart was

purchased for what was then a considerable price of AU$5000. The outlay at that time meant that Buxton held back on purchasing another work of contemporary Australian art for the next seven years.

A decade later, when Buxton was working for the development company Becton, Buxton started buying art for the Becton office. There were works by Arthur Boyd and Sidney Nolan. But the real 'thirst' for serious collecting came after Buxton saw an exhibition of a private collection owned by Loti and Victor Smorgon which was curated by John Buckley. Bequeathed to the

CONTEMPORARY AUSTRALIAN ART

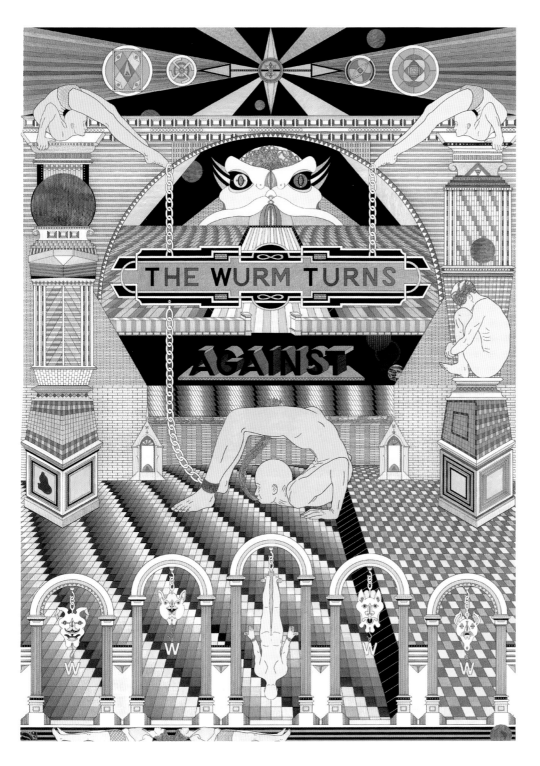

Museum of Contemporary Art in Sydney, the Smorgon's collection included works by artists Peter Booth, Davida Allen and the late Howard Arkley – the latter whom also appears in Buxton's personal collection (as pictured on page 32).

One of the other triggers for Buxton's interest in art came in the early 1990s when he enrolled to study it. Although business took Buxton away from completing the course, it gave him the opportunity to look at art, particular contemporary art, in a new light. 'That time, however brief, provided a new level of interest,' says Buxton, who focused

his attention on a few leading contemporary Australian artists. Arkley was one of those artists who struck a chord with Buxton. Arkley's *A Large House and Garden* (1997) was purchased, together with Arkley's *Freeway Exit* (1999), painted a couple of years later. 'It's the type of thing you see every day going to work. But he's taken the humble 1950s brick veneer house to a different level. They're absolutely beautiful through Arkley's paintings,' says Buxton.

Tony Clark is another artist in Buxton's collection, as are the works of John Nixon, Callum Morton, Peter Booth and photographer Bill Henson. And unlike many collectors, who purchase art or objects on the secondary market, Buxton deals directly with established galleries representing Australian contemporary art.

'The purpose from the start has been to purchase the best in Australian contemporary art,' says Buxton, whose own office is filled with great contemporary art and sculpture. There is a piece by sculptor Augustine Dall'Ava that takes pride of place on Buxton's desk. The life-size version can be found at Buxton's coastal property. There is also a painting by artist Michelle Ussher above the lounge in his office. Every corridor is also presented as a gallery, with paintings on each wall. There is a Patricia Piccinini immediately outside his office door, as well as a work by Sally Smart. Buxton is extremely proud of his Peter Tyndall artwork, *A Person Looks At A Work of Art* (2000).

While Buxton regularly goes to galleries worldwide, he also engages three art curators, namely Luisa Bosci, Business Manager of the Michael Buxton Collection; Samantha Comte; and Mark Feary. The team meets quarterly and presents a curatorial report on which artists or paintings should be collected. The curatorial team presents artists to the Board. 'The work has to be new and dynamic. It must be at the forefront of Australian contemporary art,' says Bosci, who looks at the progression each artist has made and their position with other artists represented in the various collections. 'But ultimately, the work must resonate with Michael, whatever form the art takes,' adds Bosci.

CONTEMPORARY AUSTRALIAN ART

According to Bosci, the process – as much as the finished work – can also influence whether an artist is collected. 'Michael loves seeing process. Even before Mark Hilton's *Don't Worry* (2013) was completed, Michael knew he had to have it for the collection. Michael will often collect working drawings.' Size also doesn't deter Buxton's choice, buying pieces usually only befitting an institution. 'He will always find a place, whether at home, the office or at his beach house,' says Bosci.

For Buxton, collecting the best of contemporary Australian art will occupy his time for the next 20 years. 'It is addictive, particularly as you are always seeing new work and looking at paintings that you've always wanted, but couldn't always have,' says Buxton, who recently acquired Juan Davila's *Guacolda del Carmen Gallardo* (2004). 'It's a masterpiece. What more can be said?' Few, if any, would disagree.

椅子

克里斯·康奈尔

CHAIRS

CHRIS CONNELL

DESIGNER CHRIS CONNELL has been collecting chairs since he was 16 years old. His mother had a passion for antique furniture. Initially, Connell would rummage for chairs in secondhand shops. 'I can still remember my mother spotting a Victorian-style captain's chair for the princely sum of a few dollars,' says Connell, whose first antique purchases were phones and radios. By the early 1980s, Connell was working for Kazari, a business importing Japanese furniture. 'I used to travel to Japan with the owner, Rob Joyce. He provided me with a foundation for looking at furniture,' says Connell.

Soon after, Connell opened his own store, selling furniture from the 1950s and 60s. 'I didn't look at names. I bought what 'spoke' to me,' says Connell, whose *Pepe* chair, designed by himself, features at the Museum of Modern Art in New York.

In the late 1980s, Connell purchased his first designer chair, designed by Michele De Lucchi in the Memphis style. A sculptural piece, this chair creates a welcoming sense of arrival to Connell's home, which he shares with partner and interior stylist Wendy Bannister. 'Figgins Diorama [department store] was closing down and they were selling off all the fixtures and fittings. At only AU$90, I should have bought more,' says Connell, who was attracted to the chair's

CHAIRS

form, structure and detail. 'The Memphis movement really turned things upside down, extremely unconventional,' he adds.

Since the first chair, others have followed. There is the Alias Sri Italia *Light Light* chair (1988) – a carbon fibre prototype that sits next to his bed instead of a side table – with a simple and elegant streamlined design. 'It's relatively crude in form, but it serves its function,' says Connell. Equally cherished are his two chairs designed by architect Frank Gehry – one placed in his bedroom as a side chair, the other in the living area. The timber chair in the living area, with its ribbon-like backrest, designed in 1993, complements artist Dale Frank's honey-coloured painting.

One of Connell's most-used pieces is the *Womb* chair by Eero Saarinen, designed in 1947. The moulded fibreglass seat envelops the sitter and is ideal for reading. While the couple's Siamese cat, Mr. Kitty, is allowed to sit on Connell's lap, he is discouraged from getting too comfortable or leaving fur. But he's more than welcome to join the couple when dinner is served at the dining table, with the dining chairs designed by Hans J Wegner (*Y chair*, 1950). Gio Ponti's chair, with its rattan seat and lightweight timber frame, is a decorative rather than functional piece. 'It's extremely light. You can pick it up with one finger,' says Connell.

Other chairs, such as Enzo Mari's *Box* chair – an injection-moulded polypropylene chair in vibrant blue, designed in 1975 – was conceived to be flat-packed.

One of the couple's largest pieces of furniture are the red vinyl chairs in the 1964 Fiat Sedan, restored by Connell over many years.

While the Fiat is polished and restored to the highest degree, the small timber African stool adjacent to the car is considerably more crude, with only three legs supporting a simple circular seat. 'It's quite primitive. But the three legs find their own resting place,' says Connell.

Some of Connell's chairs, including the lounge he designed personally, are used every day. Sometimes, he will sell one of his chairs and replace it with another. But he rarely goes on the 'hunt', preferring to walk into a shop or market and spot a chair that has caught his eye. 'I've never actively looked for a chair. It could be found in a shop, but sometimes, I will see a chair on the side of the road. You love it, and want to take it home, irrespective of who has designed it,' says Connell, who sees his chair collection as organic in nature – growing, as well as shrinking – in line with his tastes.

CHAIRS

SUSAN CURTIS
精美艺术品
苏珊·柯蒂斯

FINE ART

SUSAN'S THREE-LEVEL home is filled with fine art, sculpture and photography. Each wall of the house, designed by Wood Marsh Architects, tells a different story. 'We've [Susan and her late husband, Harry Curtis] been collecting art since the 1960s. We gravitated towards the same artists,' says Susan, who studied art and design, as well as having artists in the family. Her great grandfather was an artist with the Heidelberg School.

After her honeymoon, setting up a household was the next step. But rather than buying the necessary household products, they returned with an etching. 'I can still remember the expression on our parent's faces. They thought it was extravagant for a couple starting out,' says Susan. And while household items were later bought, so were more prints – some by artists as famous as Picasso.

When Susan and Harry decided to start a family, Susan gave up her full-time job as a fabric designer and, with her friend Helen O'Keefe, opened a small gallery-style shop selling prints and handcrafts. Other friends then opened a restaurant and suggested she use part of the space as a gallery, specialising in paintings. 'At that time, we exhibited emerging artists,' says Susan.

By the 1970s, Susan and Harry were becoming serious art collectors, purchasing works by Sydney Ball, Alun Leach-Jones and Fred Cress. 'We tended to buy the artists we

were supporting in the gallery,' says Susan, who went on to buy a number of works by sculptors David Wilson and Clive Murray-White. Joy Hester's works also feature on the walls. 'I still remember when one of my art teachers in [high] school came into class and spread some of Hester's work across the floor. I think one's taste in art starts early in life. Her work goes immediately to your heart,' says Susan, referring to Hester's soulful way of depicting faces, with their protruded eyes.

An exhibition of Gascoigne's work had a similarly emotional effect on the couple. Their first Gascoigne piece is displayed in a clear Perspex box next to the fireplace in their informal living area. The composition includes a roofing tile and a beer container set in a wooden box (circa 1980s). Complementing the Gascoigne work is a painting by Jennifer Joseph and a painting by Stieg Persson.

While the dining area features fine period furniture, the art is contemporary – including a painting by Sean Scully and a bitumen block print on paper by Richard Serra. There's also a lithograph by Robert Motherwell and paintings by John Walker and Jan Nelson. 'Even as a student, I was inspired by the Bauhaus movement. I love the simplicity and minimal lines,' says Susan. These carefully articulated lines are captured in Robert Owen's sculpture on the terrace leading from the dining room.

FINE ART

Although the paintings and artwork appear to have grown spontaneously in the home, Susan has a curator's eye, and is thoughtful of how work is displayed – on its own and with other artists. 'I've always been conscious of symmetry,' says Susan, pointing out the two staircases. However, art is purchased for its own qualities, not because there's a space on a wall that needs to be filled. 'I have high ceilings, with tracks on the top of each wall to give me flexibility. But each work stands on its own merit,' she adds. 'Even though there's little wall space left, if I am attracted to a certain piece of work, I'd find the space.'

Some grouping in the Curtis household shows extremely different 'brushstrokes'. For example, in the staircase between the second and third level is a heart-shaped piece by New York artist Miriam Schapiro. Next to this work is a piece on paper by Sonia Delaunay. Central to the arrangement is an Eastern-inspired painting by Tim Johnson. And a Gascoigne work, featuring a 'Slow' road sign. 'It wasn't deliberate, but there are several elements bridging the work,' says Susan, pointing out the golden halo surrounding the Buddha in Johnson's painting and the curve in Gascoigne's sign.

'It's not a conscious thing,' says Susan. In the formal living room are also several photographs by Bill Henson, Henri Cartier-Bresson, William Eggleston, Wolfgang Sievers and Grant Mudford. There are also a number of photographs by their son, Andrew Curtis, including overscaled images of a building site by the staircase.

Some art is modest in scale, while other pieces, such as a large corrugated steel sculpture in one of the studies at ground level, dominates the space.

FINE ART

FINE ART

BEATA AND VANN Fisher, both lawyers, had not really considered buying art until the turn of the millennium. It wasn't the heralding of a new century that encouraged them, but seeing the impressive contemporary art collection of a friend. 'Vann and I were transfixed by a painting by Ugo Rondinone. Really, before then, we were more than content with our small collection of paintings and artefacts, none of them really particularly valuable,' says Beata. 'At that point, you could say we were intoxicated with contemporary art and keen to learn more,' says Vann.

The couple started visiting local galleries, both in their city and interstate. Gallery owner Ray Hughes took the couple into his loft, introducing them to artists and providing them with histories and back-stories for many of the works. Other galleries made their back rooms available, as well as their extensive expertise and knowledge. The Olsen Irwin Gallery showed them the work of Amanda Marburg. 'Marburg was probably our first 'serious' piece of art,' says Beata, who now has a number of the artist's works.

Their interest for fine art – whether it was painting, sculpture or an installation – took hold. By the time they were planning the

BEATA & VANN FISHER

FIN ART

精美艺术品

贝亚特和范恩·费希尔

build of their house – designed by architect Russell Casper, then design director for Grodski Architects – nooks, walls and spaces correlated to various works of art. Rather than traditional fixed walls, the Fisher's home is not dissimilar to a warehouse gallery, with sliding walls designated for larger paintings. One such painting, by Robert Owen, features in the couple's formal dining area. In 2005, as they were travelling to purchase the Robert Owen painting (pictured on page 61, top [at left]), they read in a newspaper about the opening of a Sam Jinks exhibition. Beata can still recall her reaction to the images in the newspaper – 'Oh my god!' And so Jinks' *Pieta* (pictured on pages 56 and 60) takes pride of place in a nook in the formal lounge, specifically designed to accommodate the two realistic figures of a son holding his deceased father in his arms. 'We love his work, but we also enjoy the memory of seeing it and then purchasing it. There's also the reaction

FINE ART

from family and friends. Some who may have recently lost someone feel a little uncomfortable,' says Vann.

As lawyers, there was always an agreement between the two when the art collection was moving forward. 'We both have to love the work. If Vann sees something without me, I need to see it and agree but I can buy art without his permission. It seems to work,' says Beata. However, the works in their collection are strongly linked by colour and movement. 'We both respond to art that's arresting, literally stops you in your tracks and makes you think,' says Vann. And while some works are colourful and joyous – like Kyong Tack Hong's vibrant assortment of pencils, pens and erasers (pictured on page 63, top [at right]) – others, such as Sam Leach's paintings (see page 65), are dark and brooding.

The Fisher's art collection grew in stages – Indigenous Australian art followed by photography. The couple has several works by 'Polly' Polixeni Papapetrou as well as Callum Morton, both arranged at the entrance to the home. If they aren't familiar with a certain medium, they will start by looking at a few established artists.

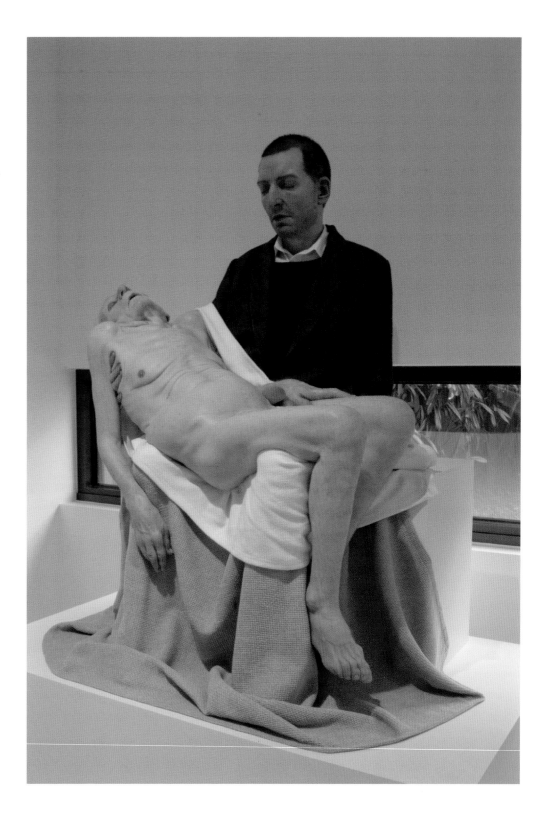

FINE ART

Sculpture is also on their radar. As well as the indoor sculptural works, such as Christopher Langton's *Give the dog a Bone* in the dining area (see page 58, bottom [at left]), there's a considerably weightier piece – a Clement Meadmore work, *Three Up* – in a courtyard. The couple still recalls the arduous process of purchasing and shipping the Meadmore piece from outside of New York. Weighing 360 kilograms, the Meadmore is an important piece in the collection.

The Fishers not only had to work out the logistics of buying something they had only seen in a book, they had to 'identify' themselves to the seller to prove that they were serious art collectors, including providing a number of references from leading galleries. Meadmore's *Three Up* also had to be collected in a relatively short time, as it was located in a public plaza and renovations were about to commence. 'We were fortunate that Russell [Casper] went over to the States to assess its condition,' says Vann.

One of the largest paintings was also 'close to never happening', at least for Vann. Darren Wardle's *Lost Weekend* (pictured on page 64), a large painting, takes pride of place in the formal living room. Vann and Beata initially saw the painting at a gallery, but there was a sold sticker next to it. The buyer was Beata, who bought it as a surprise for Vann's upcoming 60th birthday. For the next month

FINE ART

or so, Vann spoke endlessly of missing out. Even when family went to the gallery to see the painting still on the wall, with tape written below wishing him 'Happy Birthday', the 'penny still didn't drop', he says.

Since then, there have been new works as well as commissions – such as the Sam Jinks self-portrait displayed in the home office. 'We're still collecting art. I'm still keen to one day buy a painting by Ugo Rondinone,' adds Beata.

INTERIOR DESIGNER SANDY Geyer had her eyes set on jewellery as an eight-year-old girl. Her first piece was a rose gold watch, given to her by her mother. The dials of the watch are framed with bright pink enamel petals. By the late 1960s, as a student studying interior design, she frequented a jewellery store called TOR, located in a city arcade. 'In those days, I gravitated to rings,'

CONTEMPORARY JEWELLERY

当代珠宝

桑迪·盖耶

says Geyer, removing a silver ring from one of the compartments in her leather jewellery boxes. Instead of a precious stone, this ring features a series of prongs reminiscent of a toast rack. 'The store fitted with my design aesthetic,' says Geyer, who recalls missing meals to enable her to purchase another ring.

Geyer keeps many of her pieces in a series of leather bound trays, which normally sits at the bottom of her bed. There are neckpieces that feature both antique and contemporary French glass on the one strand, and rings by Georg Jensen.

SANDY GEYER

One section in the tray includes a necklace made of plastic. Another is made from wood and silver. Geyer also has a number of rings designed by Karl Fritsch, including one she refers to as her Marjorie Simpson ring – with its beehive style featuring as its crowning glory. While these precious items are beautifully kept in various sections of the trays, one's eye is drawn to the display of black mild-steel jewellery in the home's library. This display is entirely dedicated to the late Mari Funaki – who founded Gallery Funaki in the mid-1990s – an exceptional jeweller who twice received the prestigious Herbert Hoffmann Prize for jewellery.

CONTEMPORARY JEWELLERY

'When I first visited Gallery Funaki, my initial response was 'Why wasn't it already there years ago?" says Geyer, who became a close friend of Funaki's. Geyer's first acquisition of a Funaki design was a relatively simple bracelet made from black mild steel (Funaki only used this material and colour). 'I love the simplicity. There's no one else like her. If I am travelling overseas wearing one of her pieces, I'm often stopped and asked about it.' Funaki's presence on the international platform of contemporary jewellery making was unique. For example, Geyer was visiting a leading jewellery gallery

in Amsterdam and had seen a couple of rings she wished to purchase. As the gallery owner had just opened his doors, his credit facilities were not set up. 'When he saw my Funaki bracelet, he said that I could take the rings and he would send an invoice in a few days' time. The bracelet was a sign that I was a serious collector,' says Geyer.

On display are also three insect-like brooches, each wafer thin and unique. 'I bought the three at different times. They all tell a story,' says Geyer. Peter Geyer, her husband (also an interior designer), would also purchase Funaki's work for her and their two daughters on special occasions.

CONTEMPORARY JEWELLERY

While the brooches are Sandy's, there is also a piece that was commissioned by Peter as a replacement for the traditional bow tie. Funaki's mild-steel 'bow tie' clips beautifully into the buttonholes of a shirt collar. There is also a Funaki maquette on display, with a secret shelf – one of a series of maquettes that showed Funaki's talent for producing larger sculptural works. 'I call this my 'Darth Vader' bracelet,' says Geyer, picking up an angular piece.

As the two became close friends, there were many lively design discussions. Geyer was visiting Funaki's studio, which formed part of her home, and eyed a bracelet deemed a 'work in progress'. 'Mari wasn't satisfied the bracelet was going in the right direction. I suggested she add a solid form in the composition to provide more balance,' says Geyer, who now sees this piece as not only one of her most precious but, more importantly, a collaborative process.

Although the Funaki pieces are worn on a regular basis, they are kept under a Perspex case to ensure they are protected from dust. It also allows each piece to remain untouched when not being worn. 'I see these pieces not just as jewellery, but as sculpture in their own right,' says Geyer.

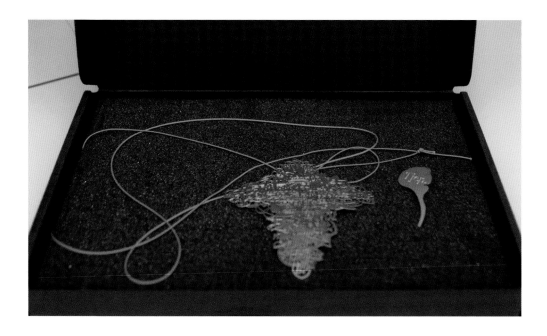

CONTEMPORARY JEWELLERY

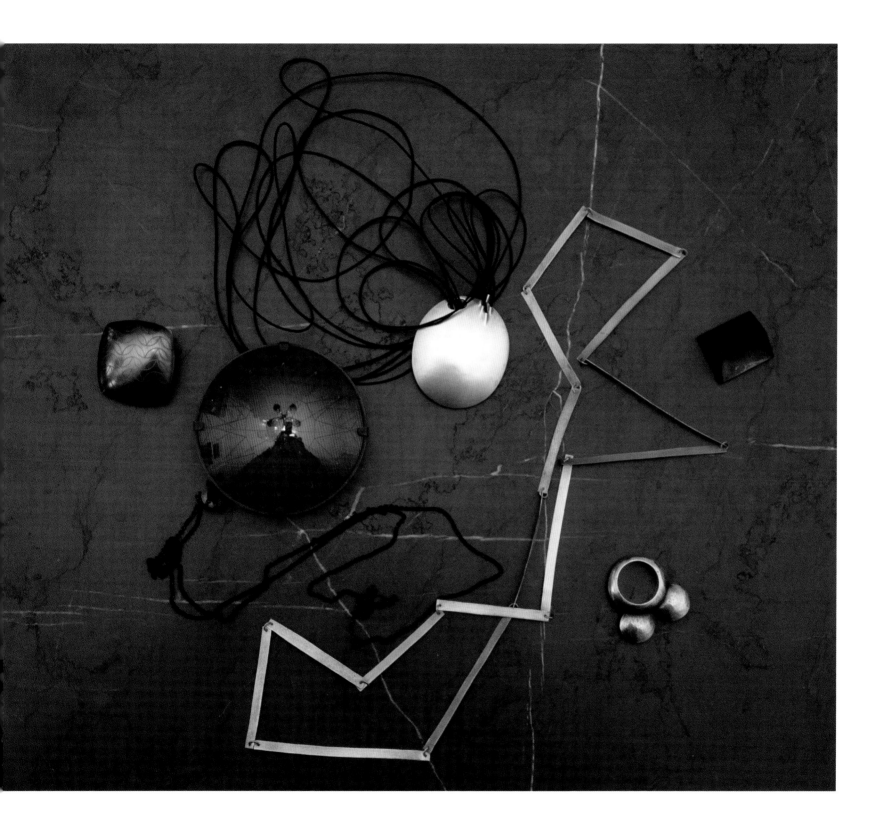

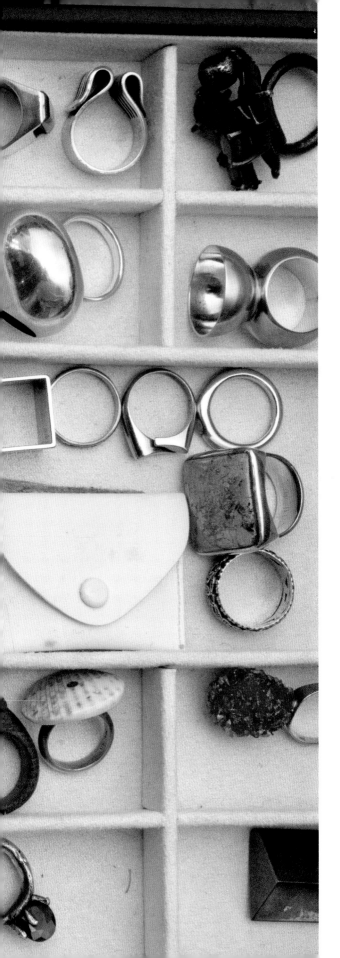

DESIGNER CHRISTOPHER GRAF lives on the outskirts of town, in a rambling Edwardian home. While the surrounding views from his terrace are impressive, it's the numerous collections inside that draw one's attention. Primitive sculptures, globe maps, candle holders, apprentice models, chandeliers, wall mirrors and vintage toys vie for space. Even signage from his former fashion boutique has been artfully arranged. Most astounding, however, is Graf's Gerry Gee ventriloquist dummy collection from the 1960s and 70s, with their wide-eyed expression that 'follows' you around a room.

Graf initially became interested in Gerry Gee dummies 20 years ago. At a toy emporium, he spotted a Gerry Gee dummy and, surprisingly, Gerry Gee's sister dummy, Geraldine. Both are now rare and are usually kept in a cupboard

VENTRILOQUIST DUMMIES
腹语式木偶
克里斯多夫·格拉夫

CHRISTOPHER GRAF

VENTRILOQUIST DUMMIES

in Graf's living room. Sometimes, they are displayed in a vintage Mercedes Benz toy car in Graf's bedroom. 'I bought both of them straight away,' says Graf, who nominates Geraldine as his favourite. 'She reminds me of Bette Davis in the film *What Ever Happened to Baby Jane?*' Wearing a tartan skirt and aqua-coloured shirt, Geraldine could equally be mistaken for her brother, except for a blonde smoke-stained wig. 'She appeals to my black sense of humour,' says Graf, who also owns silk scarves featuring images of razor blades as their centrepiece.

Perhaps due to Geraldine's expression and appearance, the dummy did not appeal to young girls when she was first released on the market in the 1960s. A few years passed before Graf gave any more thought to collecting Gerry Gee dummies or similar hybrids produced around the world.

Graf now has approximately 30 dummies that are propped upright on the bed in the guest bedroom. Like a school photo, each one smiles for the camera – the difference being that the expressions are almost identical. Not surprising, the arrangement of dummies on the bed deflects most guests from staying. 'Most people find them creepy. But I see them as part of my family,' says Graf.

Among Graf's collection is Tommy Talker, clearly identified with his name emblazoned on his jumper. Then there's Mortimer Snerd, donned in his gingham suit with red vest. His two crooked front teeth make his smirk unique, although slightly similar to his identical twin, who wears a different gingham suit. Howdy Doodey – complete with American-style cowboy boots and handkerchief – is just as cheeky with his freckled complexion.

The dummy that generally unsettles people has spiked hair and movable glass eyes. Wearing a brightly coloured patchwork jacket featuring Tobasco bottles, his eyes rarely leave you. Graf's collection also includes monocled Charlie McCartney, who wears a dapper tuxedo.

Often it's the female dummies that are the most difficult to come by. Usually, a certain dummy was created and then a female version followed. But the female dummies often did not sell and, as a result, were not produced in significant numbers. 'Many girls hated being scared by them,' says Graf. Some, such as Hobo with his red nose and tattered suit, found Graf's favour as a result of their tragic expression. 'I'm always curious about their previous lives.'

VENTRILOQUIST DUMMIES

VENTRILOQUIST DUMMIES

Graf's dummies are not about scaring people, putting off guests, or even entertaining people who visit his home. It comes from the simple pleasure of seeing amusing expressions at every turn. Sometimes, Graf's dummies are lined up on the guest bed. Other times, they can be found sitting in an armchair or driving in a toy car. 'There's no steering wheel,' says Graf, 'they're unlikely to get very far.'

WRITER SARAH GUEST enjoys looking out to her garden. But while the white camellias displayed on one wall enliven the dining room, it's her Australian timber boxes that draw one's breath. On each table, credenza and mantle piece there are beautiful timber boxes of varying sizes. Guest estimates her collection to include at least 200 boxes, most of them Australian, many dating back to the 19th century.

'My grandmother was a great collector. She was interested in British porcelain,' says Guest. 'I was continually reminded by my mother that I was like her, particularly with collecting,' she adds. As a child, Guest collected silver doll furniture from England, Belgium and France for her doll houses. This would have continued, had it not been for a move to Australia in 1961 where she raised her family.

AUSTRALIAN TIMBER BOXES
澳大利亚木箱
莎拉·格斯特

SARAH GUEST

AUSTRALIAN TIMBER BOXES

Initially, Guest set out to buy a set of drawers for her daughter. However, the secondhand shop had something else that caught her eye – a beautiful cedar box fashioned to appear as a book. Relatively inexpensive, this cedar box is designed in four parts – the top slides off as the side components are removed. 'I started to become interested in the stories about these boxes, as much as the boxes themselves,' says Guest. This cedar box, thought to be made in the early 1860s, is relatively simple compared to the other Australian timber boxes in Guest's collection. 'It was probably made by a boy as part of a woodwork project at school, or perhaps a young apprentice just starting out,' Guest says.

Guest spent the following years authoring numerous books about gardens and put the idea of collecting more Australian timber boxes out of her head. However, as she wrote more on gardens outside of the city, there were often trips to small country towns, some of which had secondhand stores and markets. One of the first boxes to catch her attention was an 1840s colonial timber box.

Made from musk and blackwood parquetry, Guest not only appreciated the design, but also the established trees felled to produce it. Relatively small in size, Guest also ponders what it might have been used for.

Other timber boxes in Guest's collection also reflect the decorative influences of the periods in which they were produced. There are the boxes created in the late 20th century, evoking the highly stylised motifs of the Art Nouveau period. The Art Deco period (1920s to late 1930s) is also represented in this collection. 'The geometric lines of these boxes can also be seen in the furniture from this period,' says Guest, who also collects presentation boxes and tea caddies.

Then there are the decorative touches. For example, one box made between 1890 and 1900 has been burnt in a manner to

depict the Australian landscape. One of Guest's most recent acquisitions is a Federation box, complete with ornate Art Nouveau-style fretwork and two mirrors. 'These mirrors would have most likely contained photographs,' says Guest. While the exterior of this box is in mint condition, the interior is described by Guest as being 'a train wreck'. The silk and wool lining is slightly shredded.

AUSTRALIAN TIMBER BOXES

Guest's collection also includes numerous 'work boxes', now known as sewing boxes, complete with threads and bobbins. Made in the late-19th century, these workboxes would often be seen next to an armchair in a drawing room 'even when visitors called', says Guest. 'It was a sign the lady of the house kept busy.'

Guest enjoys picking up her timber boxes and running her fingers along them. 'You rarely see a curved line in the design. Each piece is usually straight, except when animals are depicted.' These timber boxes are often bought for their stories as much as their fine parquetry work. There is one box thought to be from the late 1940s, when many immigrants left Europe for Australia. This box, with its metal brace-like corners, appears as a miniature suitcase.

As well as the workmanship, Guest appreciates the humour in some of her timber boxes. One is displayed on a table in the entrance to her home. This box features an Australian farm scene. As well as the settlers standing at their front door, there is the washing line, with clothes billowing in the wind. 'It's such a wonderful Australian scene,' says Guest.

AUSTRALIAN TIMBER BOXES

AUSTRALIAN TIMBER BOXES

NOLA HARGREAVES STILL recalls going secondhand shopping with her mother as a young girl. They'd rummage through dresses, shoes and bags. In the 1980s, when Hargreaves was a teenager, her focus was on crêpe dresses from the 1940s and 50s. So it seemed only natural that she opened a store selling vintage fabrics in the mid-1990s. Nowadays the store is brimming with fabrics and wallpapers from the 1940s through to the 1970s. 'Everything that I bought, I researched,' says Hargreaves, who continues to search for unique fabrics in other secondhand stores and markets. Customers also regularly drop in with a bolt of fabric, preferring to sell it on than take the scissors to it.

One of the largest finds, which formed the basis for Hargreaves' fabric collection, started when she inspected the home of a dressmaker

珍稀旧时织品

诺拉·哈格里夫斯

VINTAGE AND RARE FABRIC

NOLA HARGREAVES

who used to create dresses in the 1950s and 60s. 'When I entered her sewing room at the back of her house, there were wall-to-wall shelves of untouched fabric from that period,' says Hargreaves. 'Sekers Silks were one of the leaders at that time. I was in awe,' she adds.

Rather than spend hours on selecting individual bolts of fabric, Hargreaves purchased the lot, including classic peony patterns from the 1950s. Slowly, the collection has been increased, now including fabrics designed by Frances Burke in the 1940s and 50s, Russell Drysdale in the 1960s, and Peter Stripes in the 1970s. 'I'm afraid to wash the

VINTAGE AND RARE FABRICS

Peter Stripes cloth as the dye could run,' says Hargreaves, pointing out the unique Australian style of the wattles and wildflowers.

While many of the fabrics are organised on shelves, others are stored in crates – each one clearly identified – such as bark fabrics from the 1950s or geometrics from the 1960s. Many of the geometric patterns are inspired by the work of British artist Bridget Riley. Other fabrics are treasured for their outdated or 'incorrect' messages. For example, one fabric is covered with images of smoking packets. Other fabrics simply remind Hargreaves of a time when people dressed for the occasion. 'You can imagine how women in the 1950s found that special fabric covered in palm trees, to make a dress to wear to the airport, or the ocean liner.' Other fabrics, such as one that pictures cocktail glasses, evoke a sense of celebration – ideal for the party dress. 'These patterns still remind me of the dresses my aunties wore in the 1970s,' says Hargreaves. Then there are the one-off designs – a knit jersey with the Billboard logo from the 1980s or a length of fabric specifically designed for Melbourne's Olympic Games in 1956.

Among Hargreaves' most valuable fabrics are textiles of Frances Burke, with one of her earliest prints featuring an abstract design in pinks and cream from the late 1940s. Then there is *Craze*, a vibrant turquoise and black diamond design from the early 1950s.

One of Burke's most popular designs, often used to furnish beach houses, is the *Regency Stripe* in wide bands of blue and cream. While Hargreaves can easily identify Burke's designs, her signature and the name of the fabric is often found on the selvedge of a roll.

Some designs can be found in an entire roll of several metres, others just as pieces. In some cases, a pillow-sized swatch is all that remains. Hargreaves has a couple of Burke's dress fabrics that feature circus themes. 'Frances wasn't recognised for her dress fabrics, so these are particularly rare.'

Although the majority of fabrics in Hargreaves' collection are from the post-war period, she also has a number of dress fabrics from the early 1930s. Made of rayon and produced in America, they are displayed on shelves and tables.

One of Hargreaves' most valuable fabrics is a bolt of 11 metres, designed by artist Russell Drysdale in the 1960s for Sekers Silks. The quality and condition of this fabric is truly remarkable – as pristine as when it was first produced. 'A woman came into the store, who had had it since the 1960s. She was going to turn the fabric into drapes, but never got around to it,' says Hargreaves, who has seen this design sell at auction for around AU$2000 per metre. The fabric, in sepia tones of brown and green, depicts a series of figures in an abstract landscape. Needless to say, this fabric will remain in Hargreaves' collection, carefully protected in plastic. Although some fabrics are sold on to Hargreaves' customers, many (such as the Burke and Drysdale fabrics) will remain core to her collection. 'I love all the designs. They take to me back to my youth when I was scouring vintage fabrics with friends,' she adds.

VINTAGE AND RARE FABRICS

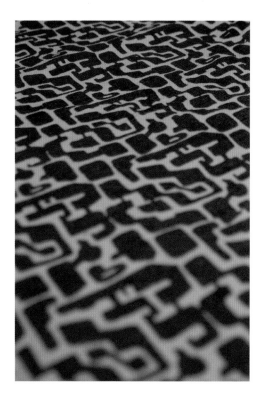

DESIGNER FURNITURE 名品家具

JOHN HENRY

约翰·亨利

ARCHITECT JOHN HENRY lives in a warehouse-style home on the edge of town. The numerous levels within the home seem to have been created with Henry's furniture collection in mind. Almost every surface features a designer chair, many of which were created by some of the world's most eminent architects – Frank Gehry, Robert Venturi, Le Corbusier and Mies van der Rohe, just to name a few. 'I'll often follow the work of leading architects who I admire. I usually find there's a great chair, lounge or table associated with their names,' says Henry, who estimates to have a collection of approximately 500 designer chairs.

Although Henry's home is spacious, the 500 chairs are also spread across his city apartment and a beach house. What can't fit into these places finds a home in a warehouse that Henry has leased. 'You could say I'm obsessed, but there are a number of steps in the process that give me pleasure: finding it, sealing the deal, and finally having it in front of me,' says Henry, who sources furniture anywhere from America to Australia, where he lives.

This obsession, as Henry refers to it, started in the early 1980s. His first purchase was Le Corbusier's *Chaise Longue*, originally designed in the late 1920s and reproduced decades later. Once the piece was placed in his living room, he realised a need for

DESIGNER FURNITURE

two of them, both covered in pony skin. 'I have never really understood contemporary sculpture, so the furniture becomes the sculpture,' says Henry.

Henry's father, Jack Henry, was a builder and introduced his teenage son to the architecture of Frank Lloyd Wright. While Wright's architecture inspired John Henry, it was the furniture that drew his attention. Henry now has a few chairs by Wright, including his barrel chairs from the 1940s. 'He [Wright] designed these chairs for the home of Herbert Johnson,' says Henry. Other mid-century chairs came onto the market, either through antique dealers or online. By the 1980s, Henry had already amassed an impressive collection of designer furniture, predominantly from the 20th century. Toward the end of that decade, Henry was looking for significant plastic designer chairs from the late 1960s to complement his growing interest in pop art. Henry's dining area is furnished with chairs by Verner Panton.

Gehry's chairs also feature prominently in Henry's collection. There is his red beaver armchair and ottoman (pictured on page 108), both made from cardboard and designed in the 1980s. There are Gehry's ribbon chairs, also made from cardboard. Neither are made for sitting on. However, most of the other chairs are there to be used. Grant and Mary Featherston's *Talking Chairs* (1967), originally designed for Canada's Expo 67, take pride of place in a study.

Although Henry has more than enough furniture to sit on, he can't help but talk about his latest acquisition. It's a Venturi lounge to go with his two Venturi chairs, bought online from a Chicago dealer.

DESIGNER FURNITURE

Unfortunately, while the lounge was in mint condition, the original floral tapestry had been removed. Undeterred, Henry stumbled across the work of designer Suzie Stanford (also featured in this book). 'Suzie will replace the light blue vinyl with a range of tea towels,' says Henry. 'I was inspired by reading Venturi's book *Learning From Las Vegas*. I've asked Suzie to find tea towels with scenes of casinos, as well as city lights,' he adds.

DESIGNER FURNITURE

Henry's collection is now a showcase – not only for architects who have designed furniture, but also designers such as Marc Newson who started his career as a jeweller. Then there are chairs by Clement Meadmore and Vico Magistretti.

Even the renowned architect Josef Hoffmann is represented in Henry's furniture collection. 'It often starts by reading about a certain architect and wanting to know more about them. Often there's a great chair in the wings,' adds Henry.

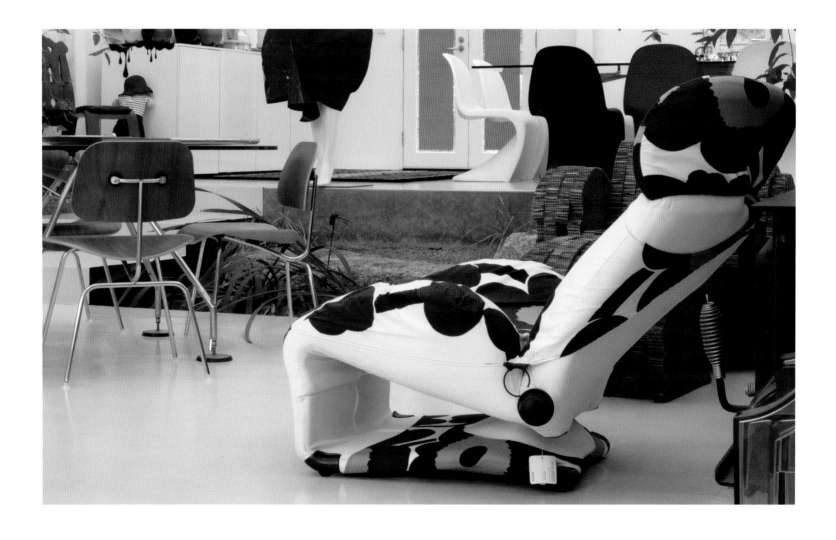

珍贵贝壳

马丁·希斯科克

MARTIN HISCOCK

PRECIOUS SHELLS

CARDIOLOGIST, DR MARTIN Hiscock started beachcombing for shells when he was just a young boy. One of his first encounters with shells was entering a shop in the Rocks area of Sydney, where there was once a business dedicated to them. 'It was overwhelming seeing all those shells, thousands of them, displayed on shelves and in drawers,' says Hiscock. As a teenager, his attention turned to diving, initially searching for cowrie shells that are native to Australia.

Fast-forward a few decades and Hiscock's knowledge and shell collection have expanded. Approximately 5000 shells are displayed in drawers, on shelves and on tables in his Miesian home designed by architect Guilford Bell. There's even a fully secured room in his home designed for the collection. Originally a sauna to accompany a tennis court, this room is now furnished with two cabinets that belonged to Paul Jones, who was a botanical arist and fellow collector of shells. Some of Jones' photographs of shells provide the perfect backdrop for the collection.

While his interest began on the beaches of Sorrento, Hiscock now contacts various dealers and specialists to add to his collection. Australian shells are well-represented in this collection and there

are also specimens from the Philippines, Caribbean and Florida. Hiscock also attends shell shows in Paris, Antwerp and the United States. While some are valued at a few hundred Australian dollars, other shells extend to five figures. Sometimes, Hiscock will buy an entire collection from another collector, even though there may only be a handful of shells of real value.

'You look for rarity, size, colour and maturity,' says Hiscock, who has drawers allocated by type of shell, be it a cowrie or wedding cake Venus shell. Shells are aligned in rows and tags are tucked into the shells to include information about each one: where it was found, who found it, and the date of collection. Others are simply too fragile to touch. But if you are able to pick one up, the detail is remarkable. For example, the 'Tent Olive' shell, originally from Mexico, features a distinctive array of tent-like shapes across its surface. The double-sided pecten shell could easily feature in a painting by Botticelli.

Then there are the thorny oyster shells, with their extraordinary spines. 'These originally came from ship wrecks off the Florida coast. The skill is in cleaning them from all the mud and coral that become attached over time,' says Hiscock.

When asked to single out his favourite type of shell in the collection, Hiscock nominates the Zoila – also one of the most sought-after

PRECIOUS SHELLS

PRECIOUS SHELLS

by collectors. 'These Zoila are now being found 100 metres below sea level by remote-operated vehicles with highly sophisticated equipment,' says Hiscock.

He also has a great interest in Polynesian art – an area where many of his shells originated. There is a fluted clamshell in the library, referred to as a bi-valve. This was purchased from a local dealer. Its miraculous form continually draws your gaze. 'It just has the most incredible spatulae. The form would probably be the envy of most architects,' he adds.

Although Hiscock has a number of shells displayed in the house, those kept in drawers are treated with 'kid gloves'. There is no direct light that would fade the shells' colour and the storage cabinets are made from wood that has been sealed. A special museum-quality material lines the drawers.

Collecting shells was extremely popular in the 19th century, particularly in aristocratic circles. Today, there is still a healthy interest in shell collecting. People like Hiscock understand and appreciate the aesthetics of these shells. 'Most people haven't seen the most beautiful shells in the world. They're locked away in museums, and only brought out on rare occasions.'

PRECIOUS SHELLS

VINTAGE SCARVES
旧时围巾
马乔里·约翰斯顿

MARJORIE JOHNSTON

MARJORIE JOHNSTON, DIRECTOR of public relations agency Wordmakers, regularly scoured secondhand stores and markets as a teenager. Her focus wasn't targeted to any specific fashion item, casting her 'net' to anything from vintage designer bags, to sunglasses or clothes. 'In the 1980s, the shops were full of clothes and accessories from the 1940s through to the 1960s,' says Johnston.

A silk georgette scarf with a 1930s Jazz Moderne pattern was one of her first acquisitions. In autumnal shades of burnt orange, mustard and burgundy, the small scarf could be worn like a kerchief. 'It reminded me of Edith Piaf, the way she wore her kerchief,' says Johnston. This scarf led her to search for others, particularly those that came with hand-rolled edges. And for her birthdays, scarves became a regular gift. 'This was one of the first scarves I received,' says Johnston, picking up a silk scarf in bands of rainbow colours, designed by Oscar de la Renta.

'By then I was hooked,' says Johnston, who started collecting scarves with more focus. There is her Hermès *Springs* scarf, which she bought new in 1992. 'I had just started in PR and one of my clients said to reward yourself from time to time when you start getting results.' In 1992, this Hermès scarf came with a price tag of AU$235, a considerable amount for someone starting out. To accompany

VINTAGE SCARVES

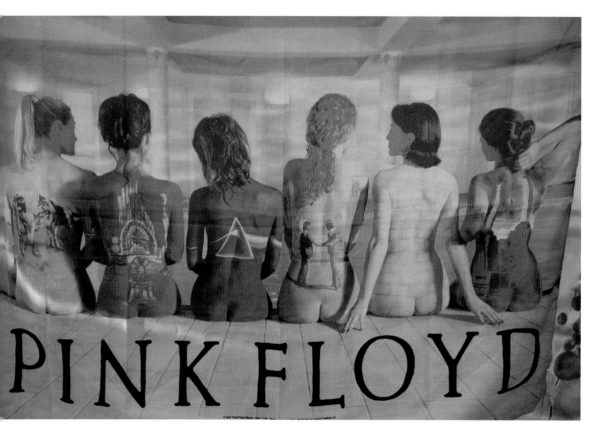

the larger purchases are the modest finds, particularly those from secondhand stores that only cost a few dollars.

Some of these scarves are now highly collectable. Her Jean de Bahrein scarf, featuring horse motifs, is now as desirable as her Christian Dior neck scarf with images of buckles and saddles. 'The scarves with horse-racing motifs are extremely popular with collectors,' says Johnston.

One of Johnston's favourite designers in her collection is British designer Richard Allan, well known among vintage scarf collectors. Some are scarves designed in the 1960s, and feature bouquets of roses. First purchased for around AU$15, a Richard Allan scarf in good condition can now fetch upwards of AU$300. 'You develop an eye for scarves, looking at the material, the way it's made and the condition it's in,' says Johnston, who avoids scarves with machine edging.

Sometimes it is the subject matter that appeals to Johnston. One of her more whimsical designs is her 'rock 'n' roll scarf'. Designed in 1987, this scarf depicts the 'back' catalogue Pink Floyd poster – featuring a series of women's bare backs printed with Pink Floyd album covers. 'It's more a curiosity piece, but it's highly valued, particularly by fans of Pink Floyd.'

Scarves don't take up much space, so finding enough has never been a constraining consideration when Johnston looks for her next scarf. Even the larger pieces, such as a woollen scarf with a rosette, can be accommodated somewhere.

Some of Johnston's scarves capture a previous era, when high-rise buildings didn't dominate the urban landscape. One of her scarves was made in Australia with the Sydney streetscape of the 1960s – featuring Kings Gate Tower and The Gazebo at Kings Cross. These city landmarks are now dwarfed by the many high-rise buildings. The only signature on the silk scarf is 'Neil'.

As well as recalling a different landscape, scarves recall movie stars, singers and models who all wore them in different ways. Marilyn Monroe wore scarves like

SYDNEY

AUSTRALIA

VINTAGE SCARVES

a kerchief, as did Piaf. Models such as Veruschka von Lehndorff, who featured regularly in magazines throughout the 1970s, often wore scarves like a bandana around her head. While Johnston appreciates the various ways scarves can be worn, she generally wears hers tied around her neck or draped over a shoulder, often worn with a jacket. 'It gives you that lift when you're seeing a client,' adds Johnston.

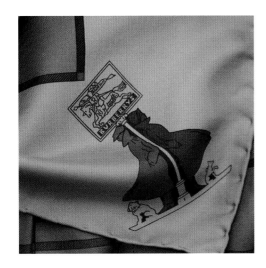

VINTAGE SCARVES

瓷器

罗谢尔·金

CERAMICS

RACHELLE KING

RACHELLE KING, a knitwear designer and collector of ceramics, can't exactly recall the date that she started collecting, but her interest started in the late 1950s when she started working in a homewares store. 'I was trying to sell my hand-painted Christmas cards, but was offered a job instead,' says King.

While the ceramics sold in the shop interested her, her first serious items were German postcards (circa 1915-17) that she found in an antique store. It was the subject matter on the cards that held her attention – flower arrangements in the most spectacular glazed vases. Some of these vases were covered in roses, while others featured serpents. And so the search began, not for more postcards, but for ceramics similar to those featured on the cards.

King found a number of amphora vases, made in the late 1800s in Austria and Germany. 'I literally fossicked through any secondhand store I passed,' says King, who found other treasures, such as ceramics by English manufacturers Shelley and Bretby. As her collection increased, shelves were either constructed or bought specifically to display them. The red shelves in the living room, floating against a red painted wall, are brimming with ceramics, primarily European. Accompanying this arrangement, are an endless number of books on collections, beautifully arranged on a glass coffee table. King delights in bringing one to attention, 'Look at this book, titled *Boring Postcards*!'

On the ground level of King's apartment is a steel display case (pictured left and right) filled with orange (or 'tango' as they are often referred to) glass vases. Predominantly Viennese glass, the colour was considered risqué at the turn of the 20th century. 'The tango [dance] gave this glass its reputation,' says King, who originally found this type of glass hidden on the top shelves of secondhand stores. But with time, and the increasing popularity of tango glass, the few pieces that can be found, are always centre-stage in the store windows. And while some collectors prefer to have one of each design, King has multiples – up to twelve in some cases.

One of King's most cherished ceramic collections is her Schafer and Vater. Displayed on a dressing table in her bedroom is a whimsical collection of these ceramics, many of which are humorous. There is the teapot, which is thought to be modelled on Queen Victoria, with her nose forming the spout. There is also a Chinese man holding a screaming child on his knees. 'Most are quite bizarre,' says King, who still loves the thrill of finding such a piece.

While the vases on each shelf aren't labelled, they have been carefully grouped. For example, next to the kitchen are shelves filled only with Awaji ceramics, produced from the 1850s on an island near Japan. 'I think I have at least 70 of these vases,' says King, who delights in the glazes as much as the unusual forms.

Around the corner in the dining room, set against 1930s glazed brick, is an arrangement of more tango glass. On the dining room buffet are Ditmar Urbach vases from the Czech Republic, produced between 1919 and 1938. There are considerably more Ditmar Urbach vases above the English dresser in the dining room, as well as on the shelves on the edge of the informal lounge.

King rarely had to spend considerable sums to purchase one of her ceramics, as they weren't popular at the time. But she clearly recalls selling her old car for AU$500, and rather than put the money in the bank, she bought her first Ditmar Urbach piece – a fruit bowl that now sits on a table in the informal living area, framed by a painting by artist Dale Hickey. Soon after, she purchased 17 more Ditmar Urbach vases.

While King adores the glazes, as well as the myriad of glass that fills her home, the irony is that she doesn't like to see flowers arranged in a vase, despite the postcards she initially fell in love with. But her passion for collecting goes on unabated. 'I don't know why people get up in the morning if they're not collecting!' she adds.

CERAMICS

CERAMICS

现当代时装

扬尼·劳福德·索尔提斯

FASHION

JANNI LAWFORD SOLTYS

1920s through to the present

IN THE EARLY 1960s, Janni Lawford Soltys was working as a model in London, after being discovered in Australia by photographer Maurice Mead (whose own collection of Clarice Cliff ceramics is also featured in this book, see page 174). Here, she not only established a modelling career, but gravitated to the many flea markets and secondhand stores in search of vintage fashion. 'I've always loved fashion from the 1930s. The beautiful tea-style dresses in silks and crêpes,' says Lawford.

'Vidal Sassoon, Mary Quant and Jean Shrimpton were part of that whole London scene,' says Lawford, who started making her own clothes before leaving Australia. She still recalls wearing short skirts, with brightly coloured tights and round John Lennon-style glasses at that time. 'The combination raised eyebrows,' she adds. And when Japanese designers became more popular in the early 1980s, Lawford embraced the asymmetrical, often oversized and deconstructed clothes. Fast-forward to the present and Lawford's wardrobe is a

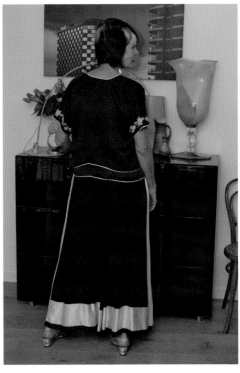

FASHION

showcase of fashion from the 1920s through to the present, regularly collaborating with designer Brighid Lehmann, who became popular in the early 1980s with her and her sister's label, Empire. 'Brighid's clothes are beautifully cut,' says Lawford, wearing one of the designer's bias-cut silk prints, evocative of Austrian artist Gustav Klimt.

Lawford still recalls visiting the Victorian market in Melbourne in the 1970s, rummaging through 1920s vintage clothing. One of her many discoveries was a red silk, three-piece suit – complete with camisole, jacket and pleated trousers. 'People weren't looking at this type of clothing,' says Lawford. In her wardrobes are also several dresses from the 1930s. One sheer, floral tea dress requires a petticoat and is teamed with Western-style boots. 'One of my hallmarks in London [1965-75] was wearing western boots,' says Lawford, who estimates she has at least 20 similar dresses in her collection.

Fashion from the late 1930s and 40s is also represented in Lawford's collection – including unworn shoes from these periods, discovered while working in New York in the 1970s. 'This shop was selling all these shoes, all unworn, still in their original boxes,' says Lawford, who restrained herself by only purchasing a couple of pairs. Consequently, one of her favourite artworks is a fabric mural from the 1940s – with her uncle, a soldier, painted on the fly of a canvas tent. 'My mother kept this work under her bed for years. She thought it was indecent,' says Lawford. While the dress she wears in the photograph of her in front of the mural (as pictured to the right) is made from 1940s fabric, it was recently fashioned into a long dress by Lehmann. 'I'm told that people often identify with the decade before they were born. I suppose my favourite period is still the 1930s – the music, fashion, as well as the art.'

By the 1950s, Lawford was searching for sundresses with nibbed waists and A-line – in silhouette. She found one such dress when visiting America on one of her modelling assignments. As with most of the clothes she wears, styles and periods are often combined in the one outfit. Earrings purchased in the 1980s are often worn with Bakelite jewellery

FASHION

from the 1920s. 'I'm not a purist. I love mixing up styles from various periods,' says Lawford. By the 1960s, Lawford was scouring out the avant-garde designers in Europe, including labels such as *VdeV* in Paris. She treasures a patchwork knitted cape she wore on the streets of London in the 1970s.

By the 1980s, Lawford was gravitating to the more streamlined, hard-edged silhouettes, discovering Lehmann's work. The black, linen pants suit, with its wide-legged trousers, is still worn – as is the asymmetrical dress from Comme des Garçons period. 'I love clothes that capture the past. I'm not a nostalgic person, but they're a continual reminder of my travels and experiences modelling in Europe.'

FASHION

澳大利亚当代艺术品

里昂住宅博物馆

AUSTRALIAN CONTEMPORARY ART

LYON HOUSEMUSEUM

ARCHITECT CORBETT LYON grew up surrounded by other architects. His mother, Marietta Perrott, was an art teacher and interior designer, regularly taking her children to art galleries. In the late 1980s, Lyon attended a retrospective of Albert Tucker's work at the National Gallery of Victoria. 'Tucker's work was spellbinding. It was telling the Australian story in a unique way – with parrots flying over antipodean heads,' says Lyon.

With some money that he had saved, Lyon was keen to purchase his first serious painting by Tucker, one of Australia's most revered artists at the time. The late Georges Mora, who then represented Tucker at his Tolarno Galleries in Melbourne, gave an important piece of advice to Lyon. 'Tucker is old and you're young. You should be collecting artists from your generation. Enjoy and follow their journey.' Mora had convinced Lyon to buy

an artist by the name of Linda Marrinon. Her latest exhibition had just opened at the Tolarno Galleries and a sold sticker was placed next to one of her key works, *Nude in a Landscape*. Over the next couple of years, Lyon started to regularly consult with Mora and his colleague Jan Minchin – who now operates Tolarno Galleries – as to which artists to look out for. 'Georges and Jan really became my mentors,' says Lyon.

Lyon's next purchase was Louise Forthun's *Orange Building Site*, a large work that captured the excavation of an impending department store. Also, well before the artist was famous, Lyon acquired several works by the late Howard Arkley, including the large mural, *Fabricated Rooms* (pictured left), and *Shadow Factories*. 'Arkley's painting talked about the suburbs, the city and the industrial landscape of Melbourne – all the things an architect is interested in,' says Lyon, who was also captivated with the artist's air-brushing technique and vivid palette. Arkley's *The Fabricated Rooms*, a 21-metre installation, features in an upstairs dining room.

AUSTRALIAN CONTEMPORARY ART

As well as Arkley, Lyon and his wife Yueji started buying the works of Patricia Piccinini. Her *Truck Babies* (pictured below) take pride of place in the living room, which is integral to the housemuseum. One of the trucks is baby pink, the other baby blue – both having their tail-ends raised in the air to give them that child-like quality. 'Patricia was just starting to show at Tolarno. It was extraordinary to see her hybrid biomorphic objects,' says Lyon, who couldn't separate the two trucks and decided, like a 'proud parent', to take home both. After this, the Lyons were committed to collecting Piccinini's work, finding space for *The Carrier* (pictured top right, opposite) – a work of a creature carrying an elderly woman – and *Atlas*, an abstract human form.

AUSTRALIAN CONTEMPORARY ART

At the Lyon Housemuseum, there is also a collection of Christopher Langton's work, including *Shooter* (pictured on page 155 [at right]) and *Swell*, and a recent collection of the artist's work that was inspired by Japanese video games. There is *Cutie* – a cartoonish female character – and *Death Warmed Up* – a male cartoon figure (both pictured here, right). Then there is *Runt* (pictured on page 155, [at left]), an over-scaled head. Just as impressive are the outdoor sculptures by Emily Floyd (as pictured on pages 162 and 163).

Pivotal to the displays is Brook Andrew's *Witness* (pictured below). Commissioned by the Lyons for the dramatic double-height space in the Housemuseum, it took a year to complete. There is an image of an atomic bomb blast off the coast of Western Australia just after the World War II, juxtaposed with a sinister-looking Turkish officer at the entrance to this imposing space.

While most of the works have found a place in the Lyon Housemuseum, two-thirds of the art is stored off site for lack of room. However, there are building plans for a new, public museum adjacent to the site. In the meantime, works are regularly changed to ensure those visiting enjoy a different experience each time. There are two works, which had been purchased before the museum was built, that have been integrated into the architecture of the space. Callum Morton's *Habitat* occupies a deep niche in the museum, while Arkley's *Fabricated Rooms* required specific dimensions. Andrew's *You've always wanted to be black (white friend)* is a permanent feature, having being painted directly onto several walls.

Lyon's decision to create a house museum started well before his extraordinary art collection took shape. Lyon had visited a number of private art collections placed in residential settings; Sir John Soane's Museum, London; The Frick Collection, New York; the Heide Museum of Modern Art, Melbourne; and the exemplary Peggy Guggenheim Collection, Venice – which Lyon first visited in 1980, the year after Guggenheim died. 'It had an amazing effect on me. You could see works by Rothko, Picasso and Pollock. All her furniture enriched the experience.'

So, in 2010, 20 years after this experience, the Lyon Housemuseum opened its doors to the public. Today, Lyon and his family come together to discuss new acquisitions. And rather than a museum that focuses on the past, it celebrates Australia's finest contemporary artists.

AUSTRALIAN CONTEMPORARY ART

DR MICHAEL MARTIN cannot put an exact figure on the number of rare books he has in his collection, estimating it to be somewhere between 1500 and 2000 books. Some are in storage, while the most prized are at arms' length. Martin's library is mainly focused on birds, but also includes early important works on plants, insects, travel and exploration. 'As a child, I was drawn to television programs on natural history,' says Martin, who was also surrounded by his late mother's portrait paintings at home. His first trip to Europe, at 20 years old, was pivotal in developing his interest in books, but also in the broader arts scene. 'I would visit galleries and museums in each city I visited.'

MICHAEL MARTIN
RARE BOOKS

书籍珍本

迈克尔·马丁

Martin carefully turns the pages of his 18th- and 19th-century works that are in original leather bindings and contain hand-painted lithographs. 'I love the experience of going into another world,' says Martin, who regularly travels to Europe to attend book fairs and visit specialist dealers.

In 1984, Martin made his first serious purchase – a three-volume set of WT Greene's *Parrots in Captivity*, published in the 1880s. Many works followed, including a spectacularly bound copy of Richard Bowdler Sharpe's *Birds of Paradise and Bower Birds*, published in the 1890s. 'I'm particularly interested in birds of paradise. Just look at the colours,' says Martin, who has personally seen 35 out of the 41 known species of bird of paradise in the wild.

At the time of pubishing these older books, there may have only been a couple of hundred copies of each produced, many of which were acquired by museums and other institutions. Although the rest have remained in private hands, many have been damaged over time. While finding a rare book always delights Martin, he is in pursuit of impeccably preserved books. 'Less is more. I'm interested in quality rather than quantity,' says Martin.

The black-billed sicklebill, *Drepanornis albertisi*, features on one of the pages and has been beautifully painted to reveal its orange- and purple-tipped wings. Another

species featured, the Huon astrapia, *Astrapia rothschildi*, is equally alluring. Each page is in mint condition. 'Foxing – a small, faded brown spot – is common in old books and occurs over time. Finding copies with little or no foxing is one of those challenges,' says Martin, who ensures that all his books are in rooms with controlled conditions.

George Dawson Rowley's *Ornithological Miscellany*, published in the 1870s, is in Martin's collection. This three-volume set, bound in elaborately gilded red Moroccan

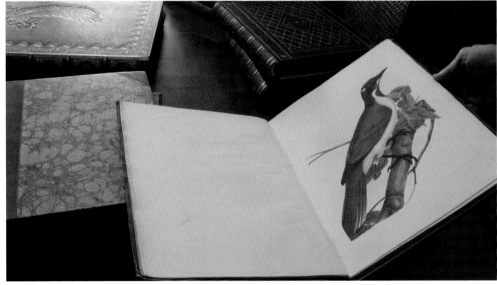

leather, is in pristine condition. There is a handwritten note by Dutch bird illustrator John Gerrard Keulemanns in one volume, confirming that he himself had hand-coloured the plates. 'My jaw hit the ground when this book arrived from England,' says Martin, pointing out the pristine marbled endpapers. For Martin, each copy must be complete and in fine original binding. 'I am always looking for the finest copies and constantly upgrade if I can find a better copy.'

Walter Buller's *A History of the Birds of New Zealand* is another title that Martin continually enjoys bringing out from the bookshelves. Published in 1873, this book appears new in its rich cobalt-blue and gilt cover. The critically endangered takahē bird is exquisitely depicted in the book and is also displayed in embossed gold relief

and time. Travel and exploration has always fascinated me,' says Martin, who spent a year working as a doctor at Casey Station in Antarctica in 1983. 'The enjoyment from exploring never ceases.'

on the cover. 'This book put Buller in the spotlight with ornithologists when his book was released,' says Martin.

Martin's collection also includes John William Lewin's *Natural History of the Lepidopterous Insects of New South Wales*, published in 1805. Each page is worth admiring for both the images and the text. While Martin attends antique book fairs and visits specialist dealers, he says collecting rare books is a solitary pursuit. 'It's not uncommon for doctors to collect books. You can look at them at night after long hours at work. It's incredibly satisfying to be transported to another place

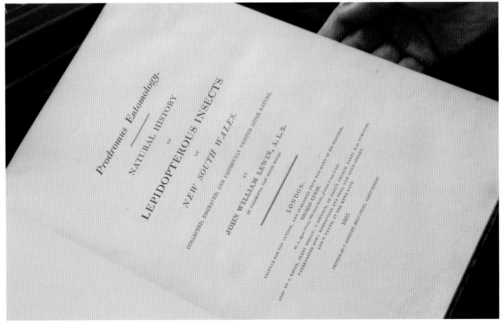

RARE BOOKS

MYIAGRA AZUREOCAPILLA, Layard.

JAN AND MAURICE MEAD first started collecting Clarice Cliff ceramics 25 years ago. They were introduced to Cliff's colourful ceramics by a friend, who had a significant collection. 'We immediately fell in love with it, both the colour and the distinctive forms,' says Maurice, who has one hand on a Cliff bowl, the other on their dog, Rufus.

While they weren't considering starting their own collection of Cliff's ceramics, they came across a small, brightly coloured vase in a secondhand shop while travelling. The abstractly patterned vase carried a price tag of AU$90. 'That was a considerable amount of money 25 years ago, particularly as it was in a junk shop,' says Maurice. Initially dismissing the vase, the couple headed back to the city. Fortunately, their daughter Jessica convinced them to turn around and buy it.

With this first piece secured, the couple had a heightened focus on Cliff ceramics when visiting other antique stores, markets and auction houses. It didn't take long before they spotted a teacup and saucer for a few dollars more. Then came a jug and a sugar bowl. One of the most prized pieces in the collection is a teapot. They had just sold a dining table at auction and bought the teapot with the proceeds. Cliff ceramics also started to become the present that Jan knew would bring great joy for Maurice's birthdays. 'Maurice was thrilled when two candle holders arrived in the post from London. Our children also bought us pieces for special occasions,' says Jan, picking up

JAN & MAURICE MEAD

克拉丽丝·克里夫陶瓷

扬和 莫里斯·米德

CLARICE CLIFF CERAMICS

one of Cliff's less flamboyant designs, a chickpot – or children's mug – made for drinking hot chocolate. 'We placed that piece in the middle of our picnic rug. We admired it for hours,' she adds.

The Mead's collection now includes vases, butter dishes, dinner plates and a set from Cliff's *Harvest* range, each with spouts and handles beautifully lined with fruit. Sheaves of wheat also add texture to each design. Many of the Cliff pieces can be identified using books on the Mead's bookshelves. Another of the Mead's most prized designs is a dinner plate, decorated with stylised Art Nouveau tree forms in a variety of deep blues.

One of Cliff's less commercially successful designs was produced in the 1920s. Believed to only have been in production for less than six months, the angular-shaped teacup (also in the Mead's collection) features a spiked surface. 'Apparently it caused blisters on the workers' hands,' says Jan.

Andrew Shapiro, director of Shapiro Auctioneers Australia, has been selling Cliff's ceramics for many years. 'One plate alone can now fetch upwards of AU$2000,' says Shapiro. He also singles out some of Cliff's designs as the most valuable. There's her *Shark's Tooth* design from the early 1920s, or the pieces that have borne the brushstrokes of the Bloomsbury Group, who worked with Cliff. 'Duncan Grant's

work is extremely rare,' says Shapiro, who sold a *Bizarre* dinner set from Grant for AU$6000 a number of years ago. 'The bright geometric patterns, such as *Lightning* from the early 1920s, are some of the most valuable,' he adds.

Most of the couple's collection is on display. Some of the vases and larger jugs are used for flowers, but it is the pleasure of walking past the collection every day that cannot be valued. 'We love the colour, the forms and that whole period of design,' says Maurice. Together, they have fine eyes for aesthetics – he being a former fashion photographer and she a model. 'We still love looking.'

CLARICE CLIFF CERAMICS

CLARICE CLIFF CERAMICS

WALLPAPERS

墙纸

PHYLLIS MURPHY

菲利斯·莫菲

ARCHITECTS JOHN AND PHYLLIS Murphy are recognised for their fine architecture after World War II. Whether as part of the team that designed the swimming pool stadium for the 1956 Melbourne Olympic Games, or their thoughtful modest homes in the suburbs of Melbourne, their name continues to be revered in architectural circles. But in the 1980s, architecture was substituted for restoring heritage buildings, as well as the collection of wallpapers that dated from the 1850s through to the late 1940s.

As early as 1958, the Murphys joined the Australian National Trust of Australia (Victoria). 'We regularly found ourselves inspecting old farm sheds with architect Robin Boyd,' says Phyllis. One of her first great attempts to protect heritage buildings came with Bacchus Manor, which was built in the 1840s and facing demolition.

While the historic home was impressive, it was the wallpapers that took her eye. 'Normally you see these wallpapers in black-and-white photos, whether taken of a grand house or in a modest worker's cottage,' says Murphy. 'From the start, I was always interested to know more about each wallpaper, how people lived and decorated their homes,' she adds.

This fascination for wallpapers increased when the couple moved to the country in the early 1980s and were introduced to a derelict painter-decorator shop from the 1850s. Even before then, Murphy collected scraps of wallpapers as part of her work with the National Trust. 'I can't recall my exact words when I saw the hundreds of wallpaper rolls stacked into shelves. They needed attention. It appeared as though children had broken in on occasion.' However, Murphy recalls seeing the vibrant colour of the papers. Forming part of a deceased estate, the owners were delighted to hand over the collection.

Once they were removed and lightly cleaned from decades of settling dust, each roll was carefully numbered and identified – a task that took months but that is now a blessing for Murphy, who can easily put her hand on the exact roll she's looking for. Many of the wallpapers come from England, France and the United States – the latter predominantly from the early part of the 20th century. Murphy's collection also includes sample books. Some, such as the chinoiserie papers, date back to the 1920s. 'I was born in the same decade. When I first started collecting papers from this period, they were simply papers I'd grown up with.'

Murphy's collection includes an extraordinary range of wallpapers, covering a variety of periods. There's the block-printed frieze from the 1860s, with classical, three-dimensional figures framed by a rich cobalt-blue sky. 'This would have graced a rather splendid dining room,' says Murphy, who carefully removes each wallpaper from its protective sleeve. More recent friezes, such as the lace-like design (1924), was conceived to be removed from its paper frame and

used under picture rails. Others have been removed in several layers – in one of the many buildings being restored.

Murphy's wallpapers provide insight into how people once lived. There is a wallpaper from the 1880s featuring a faux wood grain. 'This could easily have been from a private lounge or smoking room. It has a strong masculine feel,' says Murphy, referring to the hunted pheasant strung up by its feet. 'I love all the papers. The way each is printed on such beautiful paper and other natural fibres,' says Murphy.

WALLPAPERS

Among the collection are exquisite Art Nouveau papers and delightful stylised borders, evocative of the work of ceramicist Clarice Cliff (see the Mead collection in this book, page 174). There is also sepia-coloured wallpaper, with highly detailed floral motifs from the late 1800s. Murphy, using a paste made from flour and water, has carefully put one piece of wallpaper – smaller than an A4 sheet of paper – together. 'You can see the intensity of the blue, but also how the edges have faded over time,' says Murphy. One of the most distinctive papers is one showing a *trompe l'oeil* of a ceiling rosette from the mid-19th century. Complete with hand-painted blue ground, it would have turned heads when first applied to a Victorian ceiling.

190

WALLPAPERS

Each roll of wallpaper has been carefully labelled by Murphy with its date, maker and (where possible) its history. And while the lavish wallpapers are treasured, so are the more rudimentary papers, with their registers slightly off centre. 'Many of these papers were advanced for their time. It's not dissimilar to architecture, where there are always new developments,' adds Murphy.

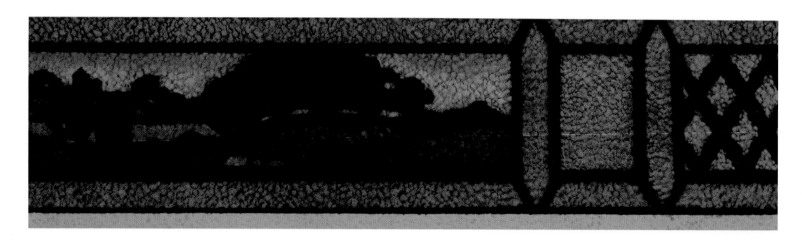

CAKE TOPPERS

蛋糕装饰

苏西·斯坦福

DESIGNER SUZIE STANFORD is a collector of many things, some of which become her furniture, lighting and bespoke objects. Her tapestries, used for chairs and ottomans, are sourced from all over the world. While several items come with fascinating stories, it is her collection of hundreds of cake toppers – figurines of wedding couples placed atop a cake – that makes it into this book. Of course, there is also Stanford's daughter, Stella, who is busy with her own collection of trolls.

Stanford, known for her 'upcycling', got the collecting itch from her grandfather. His collection consisted of tools and shells. 'The shed was a treasure trove. I'd always be asking where the shells came from and

what type of creature once lived there.' The same applied to the tools, some of them slightly worn or rusty. 'I was often attracted to the imperfections, which often told the backstory,' says Stanford.

Stanford's parents, Marian and Walter Stanford, who loved ballroom dancing and featured in Baz Luhrmann's film *Strictly Ballroom*, initiated her cake topper collection. As a five-year-old child, Stanford found her parent's own cake topper in a small box by their bedside. The small figurines of the bride and groom were wrapped in tissue paper. The effect of seeing this object was so powerful that Stanford's first question was 'When you die, instead of leaving me money, can I just have this?'

When Stanford lost her father, some of the memories of her parents' wedding day were treasured through the figurines that appeared on their wedding cake. It was this cake topper that made her start her own collection. And while Stanford's parents' cake topper is relatively subdued, her own collection is often imbued with humour.

One of Stanford's cake toppers stands alone, perhaps the result of being widowed or divorced. There is also a young man praying on his knees. Some are made from fine porcelain. Others, like her Barbie, is made from plastic and thought to have come from the 1980s. One couple continually brings

a smile to Stanford's face. Designed in the 1970s, it captures the culture and fashion of the era. The bride is wearing a dropped-waist gown and showing the signs of a baby bump, while the groom is decked out in a pale-blue suit, complete with wide lapels and flared trousers.

One of the most telling cake toppers in Stanford's collection is the before-and-after topper. From the front, the couple is at their prime, beaming with pride. From the other side, the groom supports a large potbelly and is balding. It's this humour that drives Stanford into collecting her next piece. 'When I can see a certain story,

I get extremely excited,' says Stanford, who continues to look at her purchase next to her as she drives home. Some cake toppers are purchased for a couple of dollars, while others on the 'top shelf' can cost between AU$30 and AU$40.

While the 200 pieces – often displayed in a glass cabinet in her living room – appear unique, many of her cake toppers are in fact identical, simply painted in different colours. However, the change of colour, sometimes only hair or eye colour, or in a wedding dress or suit, gives the piece its own distinctive look. Then there are her ceramic cake toppers from the 1970s, some with eyes open, others shut. 'Four of my collections are all about multiples,' adds Stanford.

CAKE TOPPERS

Even the cake toppers with misspelt quotes are treasured. Stanford loves telling the story of the reception centre owner who, in the 1970s, purchased cake toppers in volume. Unfortunately, 'Good Lock' rather than luck was written on it.

Stanford travels the world for work (many of Stanford's furniture and lighting designs are sold to leading retailers in Asia and Europe), all the while looking for clothes or art, and regularly fossicking in vintage stores and markets for certain cake toppers. Her daughter enjoys the hunt, particularly in local shops, but is as eager to build her own collection – currently standing at 276 trolls.

CAKE TOPPERS

CAKE TOPPERS

DESIGNER HATS
时尚帽子

艾莉森·沃特斯

ALISON WATERS

AS A CHILD, Alison Waters dreamed of owning a wide-brimmed hat, as worn by Madeline in the children's book of the same name by Ludwig Bemelmans. 'I still love the image of all the girls, including Madeline, lined up in two rows, wearing their hats.' Waters' passion for hats increased exponentially when she left for London in 1972, after landing a job in public relations for Biba. 'I loved going to the Rainbow Room at night, dressed head to toe in Biba,' says Waters. Some days she would dress in Marlene Dietrich-inspired gowns or suits reminiscent of Lauren Bacall, always with the appropriately matching hat.

When Waters wasn't working at Biba, she could be found fossicking for designer hats in the markets and secondhand stores. In those days, there was little competition for

secondhand Jean Muir or Chanel hats. A Christian Dior hat from the 1950s, bought at a market in Paris for a couple of hundred dollars, is now valued at over AU$2000. Another Dior hat, from the late 1940s, was purchased from a vintage hat fair and would have been designed to complement Dior's 1974 *New Look* range, in response to the liberation of the war years. Other hats have a vintage feel with a contemporary edge. Milliner Stephen Jones' raven-style hat includes a small bow dangling from its netted edge, resembling a beauty spot.

By the 1980s, Waters' hat collection included Jones, Christophe Coppens and Issey Miyake. A few Australian milliners are also included in Waters' collection, such as Gregory Ladner and Naomi Goodsir – the latter now popular in Paris. One of Goodsir's hats, called the Spanish Widow's Hat, features red bird of paradise feathers. There is also a sliver of black lace that creates a small veil over the eyes. 'You could say it allows you to start looking at other men without being too obvious,' says Waters.

Many of the hats in Waters' collection were purchases initiated by literary figures, such as Edith Sitwell – a poet from the 1930s. Then there are her hats referred to as the 'Russian' hats, all made from fur. While Waters enjoys donning these hats in winter, she also wears them in the hot temperatures of summer. 'We hold a special party around the swimming pool and everyone comes in a Russian theme. It's pure fantasy,' she adds.

DESIGNER HATS

Hats, unlike her clothing, are selected for the mood Waters is in, rather than the weather. 'I wear clothing to suit the weather. But my choice of hats is always determined by my mood. It could be whimsical or for when I feel like being more mysterious,' says Waters.

Although Waters has hats by some of the world's leading milliners, she doesn't see her hat collection as archival. Her 500 or so collection fills her inner-city apartment– stacked on shelves, displayed as art, or even found in her bath for lack of storage space. Some of the hats are packed in their original boxes, lined with acid-free tissue paper. 'I buy what I love and I wear hats every day. But I am tired of people asking if I'm off to the races well out of racing season,' says Waters, who enjoys quoting Jones' phrase that 'a hat is not merely something you put on, but is something you become.'

Waters' hat collection includes a doll-size hat, given to her by a neighbour. There is a Jones black-feathered hat, with a mischievous looking bird-like peak. 'I love hats that have a story, as well as some humour,' says Waters, picking up a hat made from a jaguar skin, previously a worn rug in the living room of her home. A black, layered tulle hat by John Rocha is as treasured as a 1950s hat found in a secondhand store while travelling to a classical music festival. And then there are the historic pieces, such as the gold and jewelled hat worn by Vivien Leigh when she came to Australia in the 1950s. 'That's another story,' says Waters.

DESIGNER HATS

版权许可
CREDITS
PERMISSIONS

All photography courtesy of Susie Latham, unless otherwise mentioned, with thanks to the following:

MICHAEL BUXTON Contemporary Australian Art (page 28)

Effigy of an Effigy with Mirage (2010), Hany Armanious, courtesy of the artist and Roslyn Oxley9 Gallery, page 29

Camouflage (2010), Pat Brassington, courtesy of the artist and ARC ONE Gallery, page 30 [top]

Untitled (2008), David Noonan, courtesy of the artist and Roslyn Oxley9 Gallery, page 30 [bottom]

The Wurm Turns Against (2014), Jess Johnson, courtesy of the artist and Darren Knight Gallery, page 31

Supa Interior (1999), Howard Arkley, ©The Estate of Howard Arkley, courtesy of Kalli Rolfe Contemporary Art, page 32

Cowboy (2007), Linda Marrinon, courtesy of the artist and Roslyn Oxley9 Gallery, page 33 [left]

Field Recording/Highland Park Hydra (2003), Ricky Swallow, courtesy of the artist and Darren Knight Gallery, Stuart Shave/Modern Art and David Kordansky Gallery, page 33 [right]

Memories of the Church of God (2010), Mike Parr, courtesy of the artist and Anna Schwartz Gallery, page 34

Shout on the hills of glory (2008), Stephen Bush, courtesy of the artist and Sutton Gallery, page 35

An Embroidery of Voids (2013), Daniel Crooks, courtesy of the artist and Anna Schwartz Gallery, page 36

Game Boys Advanced (1997-2005, 2002), Patricia Piccinini, courtesy of the artist and Tolarno Galleries, page 37 [right and left]

SUSAN CURTIS Fine Art (page 46)

Photograph of Susan Curtis, courtesy of Andrew Curtis Photography, page 46

Middle Aged Love (2002), William Kentridge, page 47 [top left]

Untitled x 2 (1980), Phillip Guston, page 47 [top right and bottom right]

Epiphany (1991), Tony Twigg, page 47 [bottom centre]

El Paso, Texas (1991), Andrew Curtis, page 48 [on wall, top left]

Untitled (1965), Tony Tuckson, page 48 [on wall, top centre]

Untitled (1970), Robert Owen, page 48 [on wall, top right]

Golden Wedge (1987), Rosalie Gascoigne, page 48 [on wall, bottom left]

Slice of Time (1979), John Firth-Smith, page 48 [on wall, bottom centre]

Untitled (1995), Craig Wise, page 48 [on wall, bottom right]

Untitled (1980), Robert Klippel, page 48 [on small table right of lounge]

Untitled (1992), Kerrie Poliness, page 48 [on far right table]

Green Tara (2005), Tim Johnson, page 49 (top left [left])

Rhythms No. 2 (1901), Sonia Delaunay, page 49 (top left [top right])

One Night Of Love (1982), Miriam Schapiro, page 49 (top left [bottom right])

Lord Street 1 and 2 (2005), Andrew Curtis, page 49 (top right [top and bottom])

Untitled (1990), John Walker, page 49 (bottom left [left])

Mind's Eye (undated), Jan Nelson, page 49 (bottom left [top centre])

Lament for Lorca (1951-52), Robert Motherwell, page 49 (bottom left [bottom centre])

Black and Red (1995), Jennifer Joseph, page 49 (bottom left [right])

Untitled (1991), Richard Serra, page 49 (bottom left [top far right])

Santa Barbara 19 (1987), Sean Scully, page 49 (bottom left [bottom far right])

Venetian Fields (circa 1975), Sydney Ball, photograph courtesy of Andrew Curtis Photography, page 49 [bottom right]

Linemarking (2013), Nick Selenitsch, page 50 [top]

Matchbox Constructions (1987-2005), Eugene Carchesio, page 50 [centre]

Calypso's Cave (1997), Ron Robertson-Swann, page 50 [bottom]

Untitled (1991), Rosalie Gascoigne, page 51 [top]

Collage (1978), Alun Leach-Jones, page 51 [bottom]

Maggie (1991), Stewart MacFarlane, page 52 [top left]

Woman With Necklace (1949), Joy Hester, page 52 [centre left]

Head (date unknown), Joy Hester, page 52 [bottom left]

Photography: G Freund, A Curtis, H Cartier-Bresson, A Kertesz, M Post-Wolcott, W Sievers, B Brandt, page 52 [right]

Hudson Industrial (1990), Jan Senbergs, photograph courtesy of Andrew Curtis Photography, page 53 [top]

Untitled (2009), Robert Owen, page 53 [bottom left]

Untitled (1983), John Dent, page 53 [bottom right]

A'dam (1978), R.H. Quaytman, page 54 (left [top left])

Vault (1982), Ron Robertson-Swann, page 54 (left [bottom right])

When Such Tenderness (1987), Victor Meertens, page 54 [top right]

Untitled (1991), Simeon Nelson, page 54 [bottom right]

New York (1975), Grant Mudford, page 55

BEATA AND VANN FISHER Fine Art (page 56)

Untitled (2004), Christopher Langton, courtesy of Tolarno Galeries, front cover (centre [left])

Circulate (2004), Christopher Langton, courtesy of Tolarno Galeries, front cover (centre [right])

Pieta (2007), Sam Jinks, courtesy of the artist and Sullivan+Strumpf, front cover [right] and pages 56 and 60

Woman and Child (2010), Sam Jinks, courtesy of the artist and Sullivan+Strumpf, page 57 [left]

Pulse #2013 (2013) Paul Snell, courtesy of the artist and Colville Gallery, page 57 [right]

Strobe Series No. 5 (2007), Marion Borgelt, courtesy of the artist and Karen Woodbury Gallery, page 58 [left]

Give the dog a Bone (2003), Christopher Langton, courtesy of Tolarno Galeries, page 58 (right [bottom left])

Pens (2013), Kyong Tack Hong, page 5 (bottom right [detail]), page 58 (right [bottom right]), page 63 (top [right])

Self Portrait (date unknown), Sam Jinks, courtesy of the artist and Sullivan+Strumpf, page 59 [left]

Sprezzatura (2011), Camille Hannah, page 59 [right]

Spectrum Shift #2 (date unknown), Robert Owen, courtesy of the artist and Arc One Gallery, page 61 (top [left])

Boulder #3 (2009), Gemma Smith, page 61 (top [right, on cabinet])

China Mary II (2002), Angela Brennan, courtesy of the artist and Roslyn Oxley9 Gallery, page 61 (bottom [left]) and page 62 (top [centre])

Tropicana (2003), Callum Morton, courtesy of the artist and Anna Schwartz Gallery, page 61 (bottom [right])

Anxiety for the Sake of Boredom (2011), Reuben Paterson, courtesy of the artist, page 62 (bottom [left])

Splash X (2014), Hung Fai, page 63 (top [left])

The Scratch (2012), Joanna Rhodes, courtesy of the artist, front cover [left] and page 63 [bottom]

Lost Weekend (2010), Darren Wardle, page 64

Nean Dasein (2007), Sam Leach, courtesy of the artist and Sullivan+Strumpf, page 65 [centre]

Standard Bearer (2007), Sam Leach, courtesy of the artist and Sullivan+Strumpf, page 65 [top right]

Chimeric Partridge (2007), Sam Leach, courtesy of the artist and Sullivan+Strumpf, page 65 [top right]

LYON HOUSEMUSEUM Australian Contemporary Art (page 154)

All images courtesy of Corbett Lyon

Runt (2011), Christopher Langton, page 155 [left]

Shooter (2011), Christopher Langton, page 155 [right]

Fabricated Rooms (1997-99), Howard Arkley, page 156 [top and bottom]

Howzat (2000), Jon Campbell, page 157

Truck Babies (1999), Patricia Piccinini, page 158

Carrier (2012), Patricia Piccinini, page 159 (top left and top right [right])

Truths Unveiled by Time #1 (2014), Daniels Crooks, page 159 (top right [left])

Cutie (Doggy Style) (2011), Christopher Langton, page 159 (bottom [right])

Death Warmed Up (2011), Christopher Langton, page 159 (bottom [left])

Witness (2012-13), Brook Andrew, page 160, page 161 (top [bottom {detail}], bottom [left {detail}])

Damien's Diamante Doggie (2008), Penny Byrne, page 6 [centre], page 161 (bottom [right])

Workshop (2012), Emily Floyd, page 162 and page 163 [bottom]

Transmutation (2010), Caroline Rothwell, page 163 [top]

致谢

ACKNOWLEDGEMENTS

STEPHEN CRAFTI

Thanks to all the collectors who appear in this book. Their drive, as well as their passion to find the next piece in their collection, is extraordinary.

My thanks also goes to photographer Susie Latham for capturing these great collections. I would also like to thank Fran Madigan for casting her critical eye over these pages.